店铺
Shop

店铺
Shop

店铺
Shop

店铺
Shop

店铺
Shop

鑫起商店

盈丰面馆

城市建筑探珍

OLD IN THE NEW

范悦 等著

中国建筑工业出版社

图书在版编目（CIP）数据

城市建筑探珍＝OLD IN THE NEW／范悦等著. —
北京：中国建筑工业出版社，2021.1（2022.3重印）
　ISBN 978-7-112-25703-4

　Ⅰ. ① 城… Ⅱ. ① 范… Ⅲ. ① 城市建筑–建筑设计–
大连 Ⅳ. ① TU984.231.3

中国版本图书馆CIP数据核字（2020）第245751号

责任编辑：徐　冉
责任校对：张惠雯

城市建筑探珍

OLD IN THE NEW

范悦　等著

*

中国建筑工业出版社出版、发行（北京海淀三里河路9号）

各地新华书店、建筑书店经销

北京锋尚制版有限公司制版

河北鹏润印刷有限公司印刷

*

开本：889毫米×1194毫米　1/20　印张：7　字数：231千字

2021年1月第一版　　2022年3月第二次印刷

定价：49.00元

ISBN 978-7-112-25703-4

（36736）

序 / Preface

范悦教授将他在大连工作期间组织的一项学术活动的成果《城市建筑探珍》的内容付之出版，并将一份打印稿交给我，问我是否愿意看一看和写个序，我说我有兴趣的是，看看时隔我执教的20世纪80年代三十多年后，范悦是如何组织教学和学术活动的。大连是个美丽的港城，也是中国近代史上留下了沉痛印记的城市。但一如上海滩上那些两个世纪以来的摩登建筑一样，作为文化结晶的物质形态的建筑又是具有超越性的人类共享的文化遗产和历史的见证，大连的近代外来建筑同样具有类似的属性。范悦在大连执教时如何对待这个城市的建筑呢？

这三十年的中国却变化得飞快，由不得你停下脚步细细思忖就一年一个样、三年大变样了，用计算机术语说就是中国按下了快进键，你没缓过神的时候那曾经的豪奢就已经是过眼烟云了。我真不知道如今老师如何带着学生对着老街讲解城市，以及如何对着老建筑开展研究和测绘。但待我打开范悦的"建筑探珍"的内容后，似乎豁然开朗。啊，好潇洒的学术活动！利用一次学术交流的档期，补上一次对城市一角的剖析。既然你无法把握结果，那就享受一下过程；既然不可能进入每家每户用钢卷尺去量尺寸，那就在它残存人间之际将最重要的、最有可能获取的重要信息研究梳理以至享受一下。于是他的"整体性认知"的新概念出来了，将城市和建筑、将过去和将来、将空间和材料等诸多要素整合起来研究一番，不求定量的分析，只求定性的判断，"看、拍、测、听、绘"放在一起做。他还设计了一个模板，归纳了八个方面让学生研究，手脑并用，且脑子要超前于手。赶在那部"视频短剧"结束之前捕捉到几个镜头，下载下几个碎片状的细部。

我仿佛看到了那一群愉快的小伙伴，在和日本学者交流和学习的过程中如何享受着自己的职业体验，如何在体味建筑细部背后的建筑师的艰辛后，又如何惊愕于城市的变化的突然，以及那背后的强力的推手。他们在享受建筑永恒的空间游戏的同时，有幸或者不幸地成为大连21世纪初城市变迁的见证人。这大约就是范悦教授这本集子的双重成果吧，是为序。

东南大学教授
朱光亚
2020年8月于石头城下

A 15　南山路杏林街交汇 / Intersection of Nanshan Road and Xinglin Street

A1　七七街 76 号 / No.76 Qiqi Street

B 19　七七街 80 号 / No.80 Qiqi Street

目录 / Contents

001 背景 / Background

002 城市建筑探珍 / Exploration of the Urban Buildings
004 大连城市由来与老建筑 / Development of Dalian and Old Buildings
008 模板解说 / Introduction of Template
010 操作方式说明 / Operation Introduction

011 图说建筑 / Graphic Expression of 47 Old Buildings

012 老建筑总平面图 / Site of the Old Buildings
014 A 区老建筑 / Site A Old Buildings
046 B 区老建筑 / Site B Old Buildings
062 C 区老建筑 / Site C Old Buildings
096 D 区老建筑 / Site D Old Buildings
108 老建筑属性 / Attribute of the Old Buildings

111 风格与构造 / Style and Structure

112 风格与构造 / Style and Structure
116 典型细部 / Typical Details
118 街区建筑 / Buildings in the Blocks

背景
Background

城市建筑探珍
Exploration of the Urban Buildings

一个关于整体性认知的建筑学记述

面对城市和建筑的快速变容，多年前我在大学里发起了"城市探珍"的活动，即走进"城市"发现"精品"。对于城市来说，作为人类"建筑"行为的历史和结果，总会蕴含着某种"记忆"或者"基因"等。发现这些基因，一方面可以帮助理解城市的由来和特质，也会更好地拓展建筑学和城市学方法的研究。

本书记述的是关于一个城市老街的建筑探珍过程，再进一步说是通过一个系统的建筑学的图示和记述模板，综合而全面地表现了老街和建筑的历史信息、场地、外观等空间信息，同时还表现了"老"建筑随时间演变的履历信息等。借助这个模板记述的系统，我们完成了南山老街47栋建筑的建筑学绘制和表述。

这里所说的建筑学的记述法，与以往的建筑测绘有很大的区别。它并不拘泥于测度的精确性，相反，而是通过模糊所谓的精准来强调要素之间的关系，重点是如何把你观察到的"建筑"整体印象和信息，直观地通过图像和图示记述下来。这种从建筑整体性出发的观察和记述，更符合人们对于建筑的认知，

也有利于纸上阅读的人比较全面生动地了解和理解"建筑"。为了实现这个系统，作者和团队尝试了从看、拍、测、听、绘等步骤来多方位记录和诠释老街的建筑历史、自发改造以及生活方式。

前面讲述的是建筑学的原理，但越过这些原理其实是一段美好的时光、一次短暂而阳光的旅程、一群愉快的小伙伴。如果没有陪同日本学者到老街探访的经历和交流，如果没有短短几周的探珍过程中遭遇突然消失的老建筑的震惊，如果没有参与其中一脸兴奋和充实的老师和同学们，这些方法和原理可能都不会出现。本书既是从建筑学原理上的对于老建筑探珍的介绍，也是纪念这段探珍经历的一个记述。

说到底，城市探珍就是对于建筑学的整体性原理的实验。每一栋建筑，它所矗立的土地、环境的历史共同形成我们认知建筑的要素，反过来说，多数这样的建筑构成对于街区绵延的印象，构成了对于城市与建筑风貌的整体性认知和理解。我们感谢参与到这个实验中的每一位老师和同学、工作室团队、堤洋树夫妇，以及参与展览和讨论的朋友、日本通世泰公司艺术画廊等。希望本书能带给快速变化的城市一点反思，带给建筑学一点启示，带给未来一点美好。

An Architectural Record about Holistic Cognition

Facing the rapid transformation of the appearance of both the cities and buildings, I launched an activity of "City-Treasure Exploration" in the university, which means searching "high-quality architectures" in the city. As the history and result of the "architectural" behavior of human beings, a city always implies a kind of "memory", or in other words, "gene". Discovering these "genes" can not only help us have a better understanding about the origin and characteristics of the city, but also enhance the research method of Architecture and Urbanology in a better way.

This book mainly records the treasure-exploration process of the Nanshan streets. Furthermore, by means of a systematic architectural graphic and record template, in a comprehensive and all sided way, we not only display the space information, including the historical information, the site and the appearance of the old streets and architectures, but also present the evolutionary information of how the "old" architectures transformed with the passage of time. With the help of the system of this template, we have finished the architectural drawings and presentations of forty-seven buildings in Nanshan Old Street.

The architectural descriptive-record method we mentioned is quite different from the traditional architecture mapping. It is not limited to the accuracy of measurement, instead, it highlights the relationship between the elements by blurring the so-called accuracy. The key point is how to intuitionally record the overall impression and information of the "architecture" you observed through pictures and graphics. This kind of observation and record is more in line with people's cognition of the architecture, and also beneficial for the people, who is reading a paper information, to understand the "architectures" in a more comprehensive and vivid way. We tried to record and interpretate the architectural history, spontaneous transformation and living style of the old streets in a lot of steps, including observing, photographing, measuring, listening, illustrating, etc.

For each architecture, the history of environment and location together form the factors of our cognition to it. In turn, most of such architectures form not only the impressions of the streets, but also the cognition urban and architectural appearances. The above description is mainly about the architectural principles, however, beyond these principles is actually a good time, a short but wonderful journey. We are very grateful to every teacher and classmate, and the team members, who participated in this experiment, as well as Mr. and Mrs. Tsutsumi, the TOSTEM Art Gallery, the friends who participated in the discussion and the exhibition, and I hope that this book can bring some reflections on the rapid-changing cities, as well as some enlightenments to the architecture, thus creating a better and wonderful future.

大连城市由来与老建筑

Development of Dalian and Old Buildings

大连是我国近代新兴的海港城市。从初建时起，先后作为俄、日的殖民地城市，进行了近半个世纪的建设。这一历史分为三个阶段，在城市建筑上反映出不同的风格特点。

1）城市起源"达里尼"

大连城市的起源始于沙俄租借时期，但当时的大连不过是一个被称为"青泥洼"的小渔村，并被起了一个具有"远东"意思的名字——"达里尼"。

沙俄统治时代的达里尼城市规划被分为三大部分，即行政区、欧罗巴区、中国人区，其中欧罗巴区又分为商业区、市民区和住宅区。面向大连湾的市内街区北侧建设了大规模的商业码头和造船厂。诸多城市广场和建筑也同期得到建设，形成了达里尼行政城市的基本面貌。

2）从达里尼到大连

1905年1月，日俄战争以俄国的失败告终，日本人将达里尼改名为大连。

日本人基本继承了俄国人的规划基础，并邀请了日本建筑师以及大批建筑大工展开建筑的修补工作和住宅建设。

南山住宅建筑以日本式的砖木结构而著称，是由日本设计师设计的，多以

北欧风格为主。它引入了西方城市地产的高效开发方式，同时创造了一种东西合璧的雅致外观。它的开放式的房屋建筑与自然浑为一体，保持和谐关系的特点。

3）"东方的巴黎"

城市建设主要是以如今的清泥洼桥一带以及人民广场周围为重点。并且向城市西部、如今的西岗区拓展。作为大连城市重要标志的火车站以及如今的市人民法院（原关东洲地方法院）等建筑，就是这个时期的代表建筑。街道从东向西舒展开来，高尔基路、凤鸣街等西部街区，那些并不高大但却和谐、并不豪华但却素雅的砖混建筑相互照应，形成街道景观。

4）气候与居住原型

东北的冬季气候比较寒冷和干燥，无法沿用日本本土的住宅样式。其中最为关键的是墙体的建筑材料，砖是保温、防火效果最佳的材料，具备了日式木结构隔墙达不到的效果。另外，双层窗、窗洞的做法也是大连特有的建筑样式。大连比较有名的近代街区有中山广场、人民广场、胜利桥北地区、南山地区、高尔基路、凤鸣街、黑石礁地区、文化街、东关街，以及著名的商业街——天津街。人们还会想起东关老街区和旅顺口区建筑群。这些街区注重人性尺度空间设计，街道高低起伏，民居和公共建筑穿插合理、错落有致。

如今这些老街区的名字依然存在，但老房子却悄然而去。老房子的保留需要街道、人文和空间的烘托，也需要将它们进行专业的绘制和记录，作为城市里时间和场景的坐标点，当人们进入其中就会自觉地唤起对于城市的记忆和认知。

【参考文献】
1. 刘长德. 大连城市规划100年 [M]. 大连：大连海事大学出版社,1999.
2. 西泽泰彦. 図说 大连都市物语[M]. 东京：河出书房新社,1999.
3. 西泽泰彦. 図说「满洲」都市物语——ハルビン·大连·瀋陽·長春 [M]. 东京：河出书房新社,1996.
4. 素素. 流光碎影 [M]. 大连：大连出版社,2008.

Dalian is a burgeoning port city in modern China, which has been undergoing a half century of construction as a colonial city of Russia and Japan since its initial construction. This historical period is divided into three stages, being reflected in different styles and characteristics in urban architecture.

1) The Origin of City "Dalny"

The city of Dalian originated from Czarist Russia leasing period. However, at that time, Dalian was just a small fishing village called "Qingniwa", and then given the name "Dalny", which means the far east.

The city planning under the rule of Czarist Russia was divided into three parts, the administrative district, the Europa district, and the Chinese district. And the Europa district was divided into commercial district, civic district and residential district. Large-scale commercial docks and shipyards were constructed on the north side of the city neighborhoods facing Dalian Bay, and at the same time, a lot of city squares and urban architectures were built, which

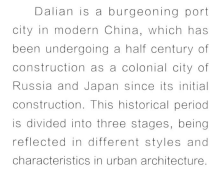

formed the basic appearance of the administrative city Dalny.

2) From "Dalny" to "Dalian"

In January 1905, the Russo-Japanese War ended in the failure of Russia, and Dalny was renamed as Dalian by the Japanese.

The Japanese basically inherited the Russian planning foundation, and invited Japanese architects, as well as a large number of skilled builders to carry out the repair work and housing constructions. Nanshan residential buildings, famous for their Japanese-style brick-wood structures, were designed by Japanese architects, mainly in the Nordic Style. They not only introduced the efficient development mode of western urban real estate, but also created an elegant appearance of the buildings, which reflected the combination of Eastern and Western elements. This kind of open-type residential building integrated with the natural environment, thus maintaining the characteristics of a harmonious relationship.

3) "The Oriental Paris"

The urban construction is mainly focus on the area around the Qingniwa bridge and the People's Square, and expand to the western part of the city (the Xigang District). The railway station and the Municipal People's Court are the representative buildings of that period and now they are also the important symbol of Dalian City. The street stretches from the east to the west such as Gaoerji road and Fengming street, and the brick buildings on both sides which are not luxurious but elegant formed the street landscape.

4) The Climate and Residential Prototypes

The winter climate in the Northeast is cold and dry, which makes it impossible to adopt the native Japanese residential style. For instance, the most critical element is the building materials of the wall. That's why they use bricks, the best thermal-insulation and fireproof materials, rather than the Japanese-style timber-structure partition wall. In addition, the design of the double window and the window openings is also a unique architectural feature of Dalian. The relatively famous modern districts of Dalian include the Zhongshan Square, the People's Square, Nanshan District, Gorky Road and so on. In addition, people will also think of the building complex in Dongguan Old Block and Lushunkou District. All of these streets and buildings pay attention to the human-scale space design. With the varying heights of different buildings, the well-arranged living space and public space blend with each other in a reasonable way.

Today, even though the names of these old streets still exist, the old buildings disappeared quietly. The preservation of the old buildings not only requires the reinforcement of the streets, humanities and space, but also needs to record and draw these building in a professional way. As the coordinate points of time and settings, people will automatically arouse their memory and cognition of the city as soon as they enter into these spaces.

模板解说

Introduction of Template

1. 照片
   Photos

   展示建筑全貌及细部

	地址 Address	建成年代 Built in	保护等级 Protection level	层数 Floors	类型 Type	用途 Use	结构 Structure	长 × 宽 × 高（m） L×W×H
35	七七街 110 号 No.110 Qiqi Street	1940 前 Before 1940	三级 Level 3	2 2	独栋住宅 Detached-house	居住 Living	砖木结构 Masonry-timber	12.3 × 8.8 × 9.5 12.3 × 8.8 × 9.5

建筑编号

2. 建筑基本属性
   Basic Attribution

   包含具体地址、建成年代、保护等级、用途、基本尺寸等信息

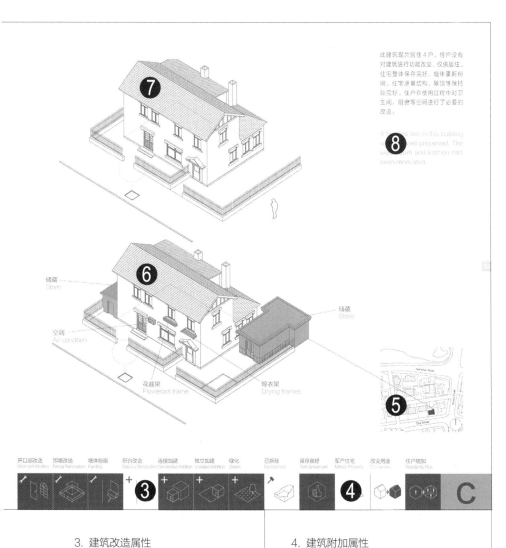

此建筑现共居住 4 户，住户没有对建筑进行功能改变，仅供居住。住宅整体保存完好，墙体重新粉刷。住宅承重结构，屋顶等保持较完好。住户在使用过程中对卫生间，厨房等空间进行了必要的改造。

4 families live in this building which is well preserved. The washroom and kitchen has been renovated.

储藏
Store

空调
Air condition

花盆架
Flowerpot frame

晾衣架
Drying frames

储藏
Store

| 开口部改造 | 围墙改造 | 墙体粉刷 | 阳台改造 | 连接加建 | 独立加建 | 绿化 | 已拆除 | 保存良好 | 军产住宅 | 改变用途 | 住户增加 | |
| Open part Worksn | Fence Renovation | Paciting | Balcony Renovation | Connective Addition | Unvided Addition | Green | Demolished | Well-preserved | Military Property | Conversion | Residents Plus | C |

3. 建筑改造属性
   Renovation Attribution
   指具体在建筑中采用的改造手法，
   标注黑色部分代表具体实施情况

4. 建筑附加属性
   Attached Attribution
   建筑改造所处状态和产权等，
   标注黑色部分代表相应属性

8. 附加说明
   Additional Notes
   包括住户信息、改造情况、
   现状描述等

7. 原状轴测
   Original Axonometric

6. 现状轴测
   Present Axonometric
   灰色：改建部分

5. 建筑总平面图
   The Site Plan

分区编号

操作方式说明
Operation Introduction

开口部改造
Door-window Renovation

更换原有窗户，提高保温性能

绿化
Green

后期使用中，庭院内种植灌木或蔬菜

围墙改造
Fence Renovation

后期改造或增加围墙

已拆除
Demolished

建筑已被拆除

墙体粉刷
Painting

通过粉刷方式，翻新建筑外观

保存良好
Well-preserved

建筑各项性能保存良好

阳台改造
Balcony Renovation

外立面增建阳台 / 将原阳台封闭，
扩大室内空间

军产住宅
Military Property

产权归属军队，居住者为军人
及家属，独具中国特色

连接加建
Connective Addition

依托原建筑加建功能房间

改变用途
Conversion

保证原有建筑形态，只改变使
用功能

独立加建
Unaided Addition

脱离原建筑物加建新的功能房间

住户增加
Residents Plus

细分内部空间，增加住户数量

注：上述图标变为反色（黑色）时，表示建筑采用此操作方式。

图说建筑

Graphic Expression of
47 Old Buildings

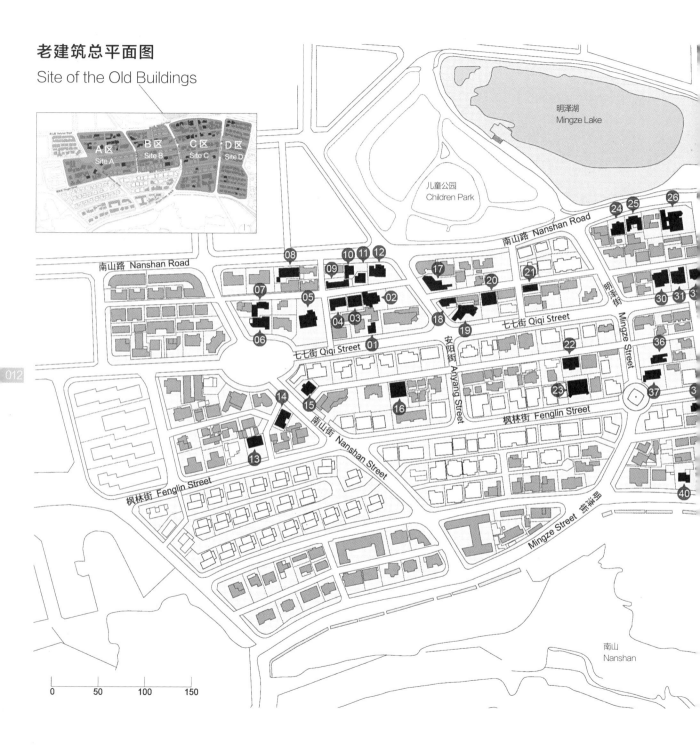

老建筑总平面图
Site of the Old Buildings

A区 Site A · B区 Site B · C区 Site C · D区 Site D

明泽湖
Mingze Lake

儿童公园
Children Park

南山路 Nanshan Road

南山路 Nanshan Road

七七街 Qiqi Street

七七街 Qiqi Street

安阳街 Anyang Street

枫林街 Fenglin Street

南山街 Nanshan Street

枫林街 Fenglin Street

明泽街

Mingze Street

Mingze Street 明泽街

南山
Nanshan

0　　50　　100　　150

A 区 / Site A

01 七七街 76 号 / No.76 Qiqi Street
02 哈尔滨街 45 号 / No.45 Harbin Street
03 哈尔滨街 43 号 / No.43 Harbin Street
04 哈尔滨街 41 号 / No.41 Harbin Street
05 南山街 233 号 / No.233 Nanshan Street
06 七七街 66 号 / No.66 Qiqi Street
07 哈尔滨街 37 号 / No.37 Harbin Street
08 南山路 165 号 / No.165 Nanshan Road
09 哈尔滨街 42/44/46/48 号 / No.42/44/46/48 Harbin Street
10 南山路 181 号 / No.181 Nanshan Road
11 哈尔滨街 50 号 / No.50 Haerbin Street
12 南山路 182 号 / No.182 Nanshan Road
13 林风街 8 号 / No.8 Linfeng Street
14 南山街 35 号 / No.35 Nanshan Street
15 南山路杏林街交汇 / Intersection of Nanshan Road and Xinglin Street
16 青林街 9 号 / No.9 Qinglin Street

B 区 / Site B

17 安阳街 80-84 号 / No.80-84 Anyang Street
18 安阳街 88 号 / No.88 Anyang Street
19 七七街 80 号 / No.80 Qiqi Street
20 七七街 82 号 / No.82 Qiqi Street
21 七七街 86 号 / No.86 Qiqi Street
22 青林街 35 号 / No.35 Qinglin Street
23 青林街凡尔赛会馆 / Versailles Club
24 南山路 207 号 / No.207 Nanshan Road

C 区 / Site C

25 南山路 209 号 / No.209 Nanshan Road
26 南山路 215 号 / No.215 Nanshan Road
27 南山路 217 号 / No.217 Nanshan Road
28 南山路 221/223 号 / No.221/223 Nanshan Road
29 哈尔滨街 76 号 / No.76 Harbin Street
30 七七街 98-1 号 / No.98-1 Qiqi Street
31 七七街 100 号 / No.100 Qiqi Street
32 七七街 102 号 / No.102 Qiqi Street
33 七七街 104 号 / No.104 Qiqi Street
34 七七街 108 号 / No.108 Qiqi Street
35 七七街 110 号 / No.110 Qiqi Street
36 明泽街 88 号 / No.88 Mingze Street
37 枫林街 36 号 / No.36 Fenglin Street
38 五五路 97 号 / No.97 Wuwu Road
39 枫林街 37 号 / No.37 Fenglin Street
40 望海街 36 号 / No.36 Wanghai Street
41 南山路 230 号 / No.230 Nanshan Road

D 区 / Site D

42 南山路 233 号 / No.233 Nanshan Road
43 南山路 201 号 / No.201 Nanshan Road
44 高阳路 41 号 / No.41 Gaoyang Road
45 林景街 2 号 / No.2 Linjing Street
46 林景街 9 号 / No.9 Linjing Street
47 望海街 59 号 / No.59 Wanghai Street

01	地址 Address	建成年代 Built in	保护等级 Protection Level	层数 Floors	类型 Type	用途 Use	结构 Structure	长×宽×高（m） L x W x H
	七七街 76 号 No.76 Qiqi Street	1922-1945 1922-1945	三级 Level 3	2 2	独栋住宅 Detached House	居住 Living	砖木结构 Masonry-timber	10.7×8.0×9.9 10.7×8.0×9.9

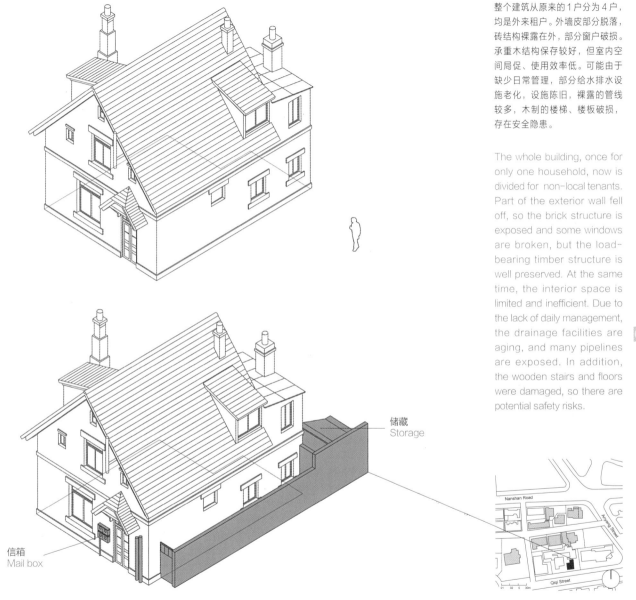

整个建筑从原来的1户分为4户，均是外来租户。外墙皮部分脱落，砖结构裸露在外，部分窗户破损。承重木结构保存较好，但室内空间局促、使用效率低。可能由于缺少日常管理，部分给水排水设施老化，设施陈旧，裸露的管线较多，木制的楼梯、楼板破损，存在安全隐患。

The whole building, once for only one household, now is divided for non-local tenants. Part of the exterior wall fell off, so the brick structure is exposed and some windows are broken, but the load-bearing timber structure is well preserved. At the same time, the interior space is limited and inefficient. Due to the lack of daily management, the drainage facilities are aging, and many pipelines are exposed. In addition, the wooden stairs and floors were damaged, so there are potential safety risks.

015

储藏
Storage

信箱
Mail box

Nanshan Road

Anying Street

Qiqi Street

开口部改造
Door and Window

围墙改造
Fence Renovation

墙体粉刷
Painting

阳台改造
Balcony Renovation

连接加建
Connective Addition

独立加建
Unaided Addition

绿化
Green

已拆除
Demolished

保存良好
Well-preserved

军产住宅
Military Property

改变用途
Conversion

住户增加
Residents Plus

ARMY

A

中华人民共和国成立后曾有高级官员居住于此。现居住 13 人。建筑的一部分作为垃圾收购站使用。由于年代久远加之使用不善，建筑外墙皮部分脱落，但内部结构稳固，给水排水设施仍可继续使用。

After founding of people's Republic of China, there were senior officers living here. Now there are 13 residents. Part of the building is used as a waste recycling station. Because the building is age-old, some parts of the exterior walls fell off. However, the interior structure is still firm, and the drainage facilities are still in good condition.

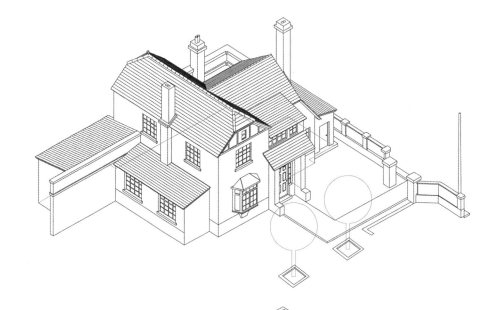

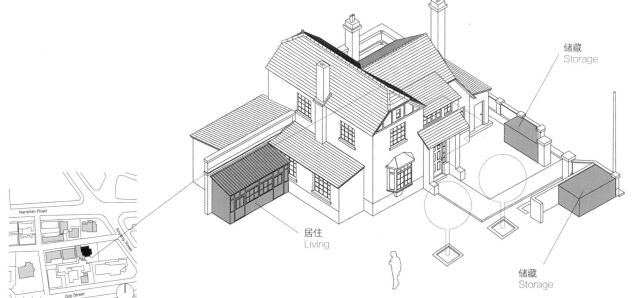

储藏
Storage

居住
Living

储藏
Storage

Nanshan Road

Anping Street

Qiqi Street

01 02 30m

02	地址 Address	建成年代 Built in	保护等级 Protection Level	层数 Floors	类型 Type	用途 Use	结构 Structure	长 X 宽 X 高（m） L x W x H
	哈尔滨街 45 号 No.45 Harbin Street	1922-1945 1922-1945	二级 Level 2	2 2	独栋住宅 Detached House	居住 Living	砖木结构 Masonry-timber	20.4×10.4×9.0 20.4×10.4×9.0

Overlook

Detail

Detail

Left elevation

开口部改造
Door and Window

围墙改造
Fence Renovation

墙体粉刷
Painting

阳台改造
Balcony Renovation

连接加建
Connective Addition

独立加建
Unaided Addition

绿化
Green

已拆除
Demolished

保存良好
Well-preserved

军产住宅
Military Property

改变用途
Conversion

住户增加
Residents Plus

A

Detail

Detail

Perspective

03	地址 Address	建成年代 Built in	保护等级 Protection Level	层数 Floors	类型 Type	用途 Use	结构 Structure	长×宽×高（m） L×W×H
	哈尔滨街 43 号 No.43 Harbin Street	1922-1945 1922-1945	二级 Level 2	3 3	集合住宅 Collective Housing	居住 Living	砖混结构 Masonry-Concrete	14.0×13.5×12.7 14.0×13.5×12.7

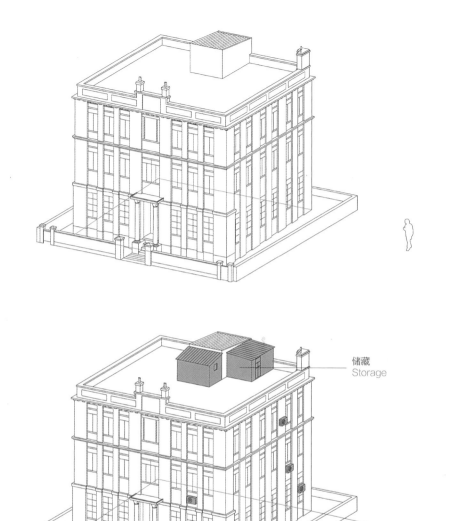

储藏
Storage

建设初期为捷克领事馆，后改为住宅，共居住7户，其中5户为外来租户。建筑体形简洁，立面采用横向三段式。外墙石材部分保存完整，局部有修补痕迹，面砖部分保存完好。雨水管外露。内部结构破坏较小，但木构及上下水等设施陈旧。

This building was used as the Czech consulate at the very beginning, and later on it was converted to a dwelling. Generally, the shape of the building is restrained with a symmetrical layout of three lateral zones. Part of the exterior-wall stone is well preserved, and at the same time, there are repairing traces on some parts of the walls. The rain-water pipes are exposed. In addition, the interior structure is less damaged, but the wooden structure and the water supply and drainage facilities are old.

开口部改造	围墙改造	墙体粉刷	阳台改造	连接加建	独立加建	绿化	已拆除	保存良好	军产住宅	改变用途	住户增加
Door and Window	Fence Renovation	Painting	Balcony Renovation	Connective Addition	Unaided Addition	Green	Demolished	Well-preserved	Military Property	Conversion	Residents Plus

A

Back elevation

Detail

Detail

Detail

04	地址 Address	建成年代 Built in	保护等级 Protection Level	层数 Floors	类型 Type	用途 Use	结构 Structure	长×宽×高（m） L×W×H
	哈尔滨街 41 号 No.41 Harbin Street	1922–1945 1922–1945	三级 Level 3	1 1	独栋住宅 Detached House	居住 / 店铺 Living / Shop	砖木结构 Masonry-timber	17.1×12×6.9 17.1×12×6.9

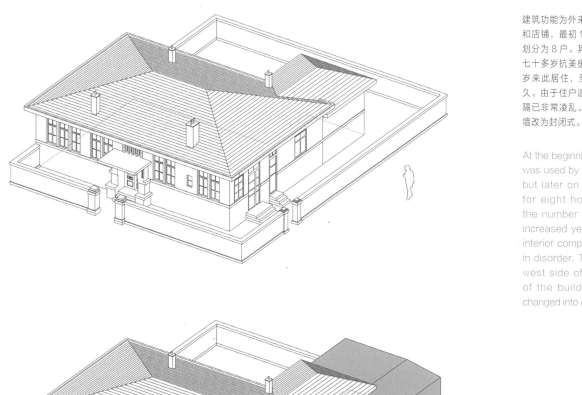

建筑功能为外来务工人员的住房和店铺，最初1户人家使用，现划分为8户。其中一户居住者为七十多岁抗美援朝战士，三十多岁来此居住，至今已有48年之久。由于住户逐年增加，内部分隔已非常凌乱。住宅一层西侧围墙改为封闭式。

At the beginning, this building was used by one household, but later on it was divided for eight households. As the number of households increased year by year, the interior compartmentation is in disorder. The wall on the west side of the first floor of the building has been changed into closed type.

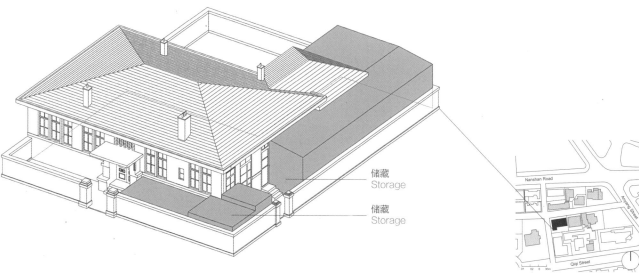

储藏
Storage

储藏
Storage

开口部改造	围墙改造	墙体粉刷	阳台改造	连接加建	独立加建	绿化	已拆除	保存良好	军产住宅	改变用途	住户增加
Door and Window	Fence Renovation	Painting	Balcony Renovation	Connective Addition	Unaided Addition	Green	Demolished	Well-preserved	Military Property	Conversion	Residents Plus

A

	地址 Address	建成年代 Built in	保护等级 Protection Level	层数 Floors	类型 Type	用途 Use	结构 Structure	长×宽×高（m） L×W×H
05	南山街 233 号 No.233 Nanshan Street	1952 1952	一级 Level 1	3 3	公建 Public Building	幼儿园 kindergarten	砖木结构 Masonry-timber	14.1×8.0×6.3 14.1×8.0×6.3

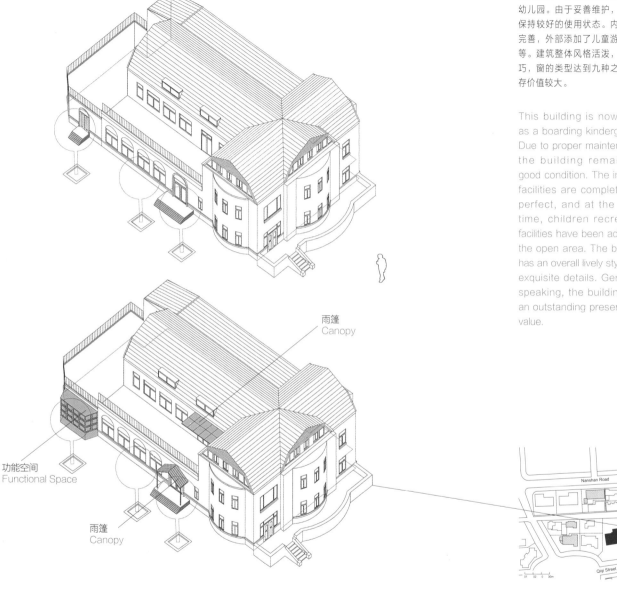

该建筑为一所集体性质的寄宿制幼儿园。由于妥善维护，建筑仍保持较好的使用状态。内部设施完善，外部添加了儿童游乐器械等。建筑整体风格活泼，细部精巧，窗的类型达到九种之多。保存价值较大。

This building is now used as a boarding kindergarten. Due to proper maintenance, the building remains in good condition. The internal facilities are complete and perfect, and at the same time, children recreation facilities have been added in the open area. The building has an overall lively style with exquisite details. Generally speaking, the building has an outstanding preservation value.

023

雨篷
Canopy

功能空间
Functional Space

雨篷
Canopy

Nanshan Road

Qiqi Street

开口部改造	围墙改造	墙体粉刷	阳台改造	连接加建	独立加建	绿化	已拆除	保存良好	军产住宅	改变用途	住户增加
Door and Window	Fence Renovation	Painting	Balcony Renovation	Connective Addition	Unaided Addition	Green	Demolished	Well-preserved	Military Property	Conversion	Residents Plus

A

Front elevation

办公
Office

Nanshan Road

Qiqi Street

0 1 0 2 0 30m

06	地址 Address	建成年代 Built in	保护等级 Protection Level	层数 Floors	类型 Type	用途 Use	结构 Structure	长×宽×高（m） L×W×H
	七七街 66 号 No.66 Qiqi Street	1922-1945 1922-1945	一级 Level 1	2 2	独栋住宅 Detached House	居住 Living	砖木结构 Masonry-timber	26.1×12.9×7.7 26.1×12.9×7.7

建筑一层为办公，二层为住宅。建筑外墙皮部分脱落，有些墙体出现纵向裂纹。

The first floor of this building is for office use, and the second floor is for living. Some part of exterior wall peeled off. There are some longitudinal cracks on the surface of part of the walls.

Front elevation

Detail

Detail

Detail

开口部改造	围墙改造	墙体粉刷	阳台改造	连接加建	独立加建	绿化	已拆除	保存良好	军产住宅	改变用途	住户增加
Door and Window	Fence Renovation	Painting	Balcony Renovation	Connective Addition	Unaided Addition	Green	Demolished	Well-preserved	Military Property	Conversion	Residents Plus

A

Left elevation

Detail

Right elevation

Back elevation

07	地址 Address	建成年代 Built in	保护等级 Protection Level	层数 Floors	类型 Type	用途 Use	结构 Structure	长×宽×高 (m) L×W×H
	哈尔滨街 37 号 No.37 Harbin Street	1922-1945 1922-1945	二级 Level 2	2 2	独栋住宅 Detached House	居住 Living	砖木结构 Masonry-timber	13.6×10.2×7.0 13.6×10.2×7.0

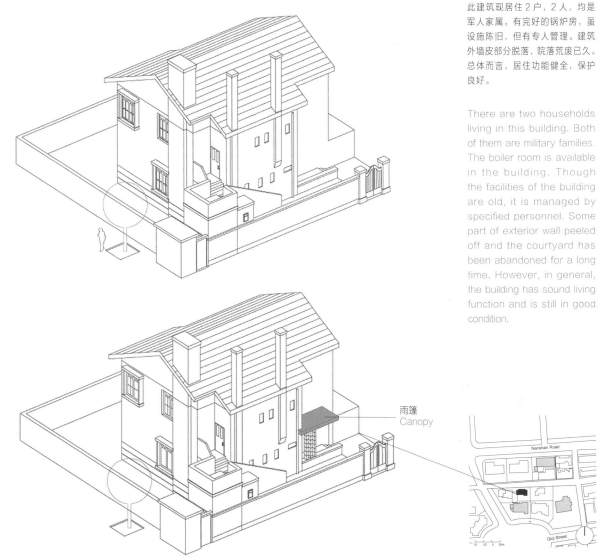

此建筑现居住 2 户，2 人，均是军人家属。有完好的锅炉房，虽设施陈旧，但有专人管理。建筑外墙皮部分脱落，院落荒废已久。总体而言，居住功能健全，保护良好。

There are two households living in this building. Both of them are military families. The boiler room is available in the building. Though the facilities of the building are old, it is managed by specified personnel. Some part of exterior wall peeled off and the courtyard has been abandoned for a long time. However, in general, the building has sound living function and is still in good condition.

雨篷
Canopy

开口部改造	围墙改造	墙体粉刷	阳台改造	连接加建	独立加建	绿化	已拆除	保存良好	军产住宅	改变用途	住户增加
Door and Window	Fence Renovation	Painting	Balcony Renovation	Connective Addition	Unaided Addition	Green	Demolished	Well-preserved	Military Property	Conversion	Residents Plus

A

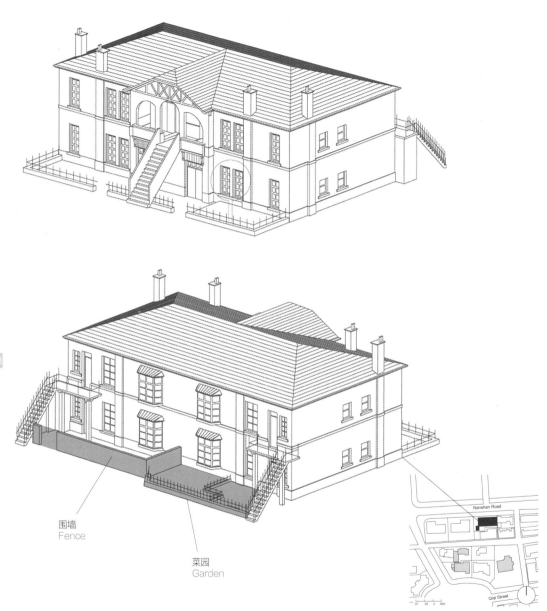

围墙
Fence

菜园
Garden

Nanshan Road

Qiqi Street

01 02 03   30m

08	地址 Address	建成年代 Built in	保护等级 Protection Level	层数 Floors	类型 Type	用途 Use	结构 Structure	长×宽×高（m） L×W×H
	南山路 165 号 No.165 Nanshan Road	1922-1945 1922-1945	二级 Level 2	2 2	集合住宅 Collective Housing	居住 Living	砖木结构 Masonry-timber	19.6×12.5×6.8 19.6×12.5×6.8

该建筑共租住 7 户 20 余人。一些雨水管破损，设施陈旧，户内上下水设施不完善，无统一管理，外部楼梯已弃用。建筑外墙皮部分脱落，但结构完好。

There are now 7 households, more than 20 people living in the building. Some rain-water pipes are damaged and the facilities of the building are old. At the same time, the water supply and drainage facilities and not complete. There is no unified management of the building. Some part of exterior wall peeled off, while the structure is well preserved.

029

Detail

Back Elevation

Detail

Detail

开口部改造	围墙改造	墙体粉刷	阳台改造	连接加建	独立加建	绿化	已拆除	保存良好	军产住宅	改变用途	住户增加
Door and Window	Fence Renovation	Painting	Balcony Renovation	Connective Addition	Unaided Addition	Green	Demolished	Well-preserved	Military Property	Conversion	Residents Plus

A

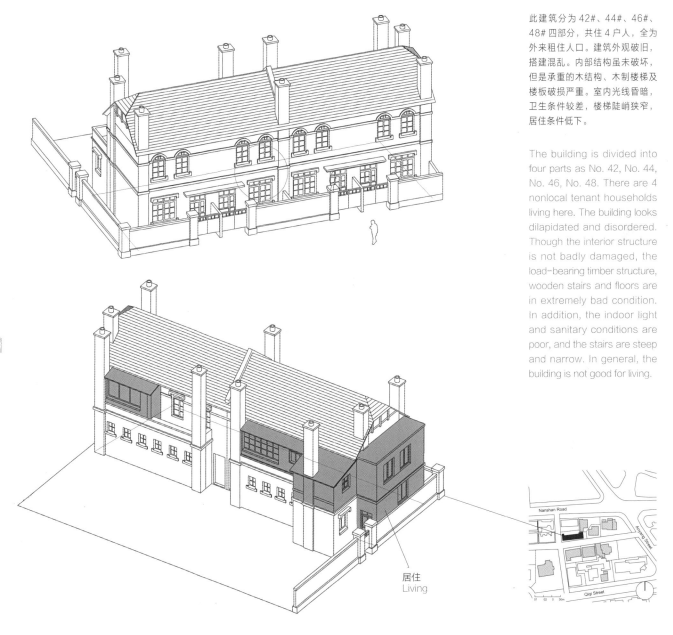

此建筑分为 42#、44#、46#、48# 四部分，共住 4 户人，全为外来租住人口。建筑外观破旧，搭建混乱。内部结构虽未破坏，但是承重的木结构、木制楼梯及楼板破损严重。室内光线昏暗，卫生条件较差，楼梯陡峭狭窄，居住条件低下。

The building is divided into four parts as No. 42, No. 44, No. 46, No. 48. There are 4 nonlocal tenant households living here. The building looks dilapidated and disordered. Though the interior structure is not badly damaged, the load-bearing timber structure, wooden stairs and floors are in extremely bad condition. In addition, the indoor light and sanitary conditions are poor, and the stairs are steep and narrow. In general, the building is not good for living.

居住
Living

Nanshan Road
Anning Street
Qiqi Street

01 02 30m

	地址 Address	建成年代 Built in	保护等级 Protection Level	层数 Floors	类型 Type	用途 Use	结构 Structure	长×宽×高（m） L×W×H
09	哈尔滨街 42/44/46/48 号 No.42/44/46/48 Harbin Street	1922-1945 1922-1945	三级 Level 3	2 2	集合住宅 Collective Housing	居住 Living	砖木结构 Masonry-timber	21.2×9.0×8.5 21.2×9.0×8.5

Detail

开口部改造
Door and Window

围墙改造
Fence Renovation

墙体粉刷
Painting

阳台改造
Balcony Renovation

连接加建
Connective Addition

独立加建
Unaided Addition

绿化
Green

已拆除
Demolished

保存良好
Well-preserved

军产住宅
Military Property

改变用途
Conversion

住户增加
Residents Plus

A

Detail

Detail

Detail

10	地址 Address	建成年代 Built in	保护等级 Protection Level	层数 Floors	类型 Type	用途 Use	结构 Structure	长×宽×高（m） L×W×H
	南山路 181 号 No.181 Nanshan Road	1922 前 Before 1922	三级 Level 3	2 2	独栋住宅 Detached House	居住 Living	砖木结构 Masonry-timber	13.2×7.8×6.8 13.2×7.8×6.8

曾有位任职于辽宁师范大学的、来自苏联的教师在此居住，现在居住着3户人家，辽师大退休老教授的妻子及女儿一家，以及一户外来租住的母子。建筑外墙无粉刷，内部结构坚实，木质楼梯及楼板维护较好。厨卫设施陈旧，3户共用一个卫生间，为正常生活带来诸多不便。

There was once a Soviet professor living here. Now there are 3 households, an old couple of professor retired from Liaoning Normal University, their daughter's family, and a tenant with mother and son. The exterior walls of the building are unpainted, and the internal structure is solid. The wooden stairs and floors are well preserved, while the kitchen and toilet facilities are timeworn. Three households share one bathroom, which causes a lot of inconvenience.

033

储藏
Storage

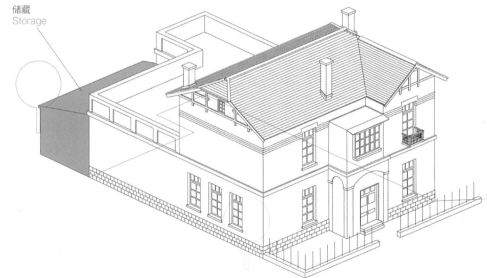

 开口部改造
Door and Window

 围墙改造
Fence Renovation

墙体粉刷
Painting

 阳台改造
Balcony Renovation

 连接加建
Connective Addition

 独立加建
Unaided Addition

 绿化
Green

 已拆除
Demolished

 保存良好
Well-preserved

 军产住宅
Military Property

 改变用途
Conversion

 住户增加
Residents Plus

A

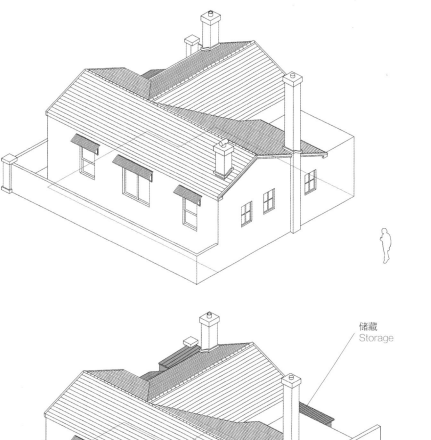

储藏
Storage

储藏
Storage

居住
Living

一栋住宅内分成 2 户，共居住 6 人，其中 1 户为老年夫妇，另外一户为外来租户。内部空间在原基础上稍作调整，北向加建花房并作为老年夫妇的入口门厅。老年夫妇住宅为 1 室 1 厅，1 套厨卫；另 1 户为 1 居室，1 套厨卫。内部进行了简单装修，利用建筑外墙与院墙搭建带状储藏空间，并增设 1 户水暖电配套设施。建筑西向和南向外墙保留原凸凹肌理和颜色，东向和北向粉刷为白色。主人强烈要求对其保留，自愿出资修缮。

There are two households, 6 people living here now. The interior space was slightly adjusted on the original basis, with a greenhouse built northward as an entrance hall. The west and south facades of the building retain the original concavo-convex texture and color, and the east and north facades were painted white. The owners strongly required the preservation of the old walls, and they are willing to pay for the renovation.

	地址 Address	建成年代 Built in	保护等级 Protection Level	层数 Floors	类型 Type	用途 Use	结构 Structure	长 × 宽 × 高 (m) L × W × H
11	哈尔滨街 50 号 No.50 Haerbin Street	1922-1945 1922-1945	三级 Level 3	1 1	独栋住宅 Detached House	居住 Living	砖木结构 Masonry-timber	11.4 × 11.4 × 4.8 11.4 × 11.4 × 4.8

Roof

Back elevation

Back elevation

开口部改造
Door and Window

围墙改造
Fence Renovation

墙体粉刷
Painting

阳台改造
Balcony Renovation

连接加建
Connective Addition

独立加建
Unaided Addition

绿化
Green

已拆除
Demolished

保存良好
Well-preserved

军产住宅
Military Property

改变用途
Conversion

住户增加
Residents Plus

A

Entrance

Window

Window

Window　Door in the yard

	地址 Address	建成年代 Built in	保护等级 Protection Level	层数 Floors	类型 Type	用途 Use	结构 Structure	长×宽×高（m） L×W×H
12	南山路 182 号 No.182 Nanshan Road	1922-1945 1922-1945	二级 Level 2	2 2	独栋住宅 Detached House	居住 / 店铺 Living / Shop	砖木结构 Masonry-timber	17.6×13.4×9.8 17.6×13.4×9.8

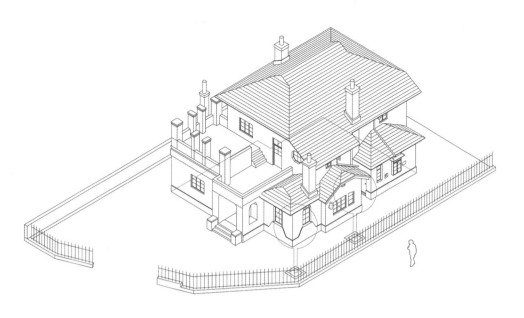

该建筑现在一层东侧被改造成汽车轮胎店，东侧院子用以停放被修理车辆。一层后院和二层为住宅，现居6户。房子保温性能良好，居住较为舒适，因住人数增加，已造多间厨房和卫生间。

The east of the first floor of the building has been converted into a tire shop in order to make it convenient to use the east yard for parking. The backyard of the first floor and the second floor of the building are used for living. The building has a good heat preservation performance, which is comfortable for living.

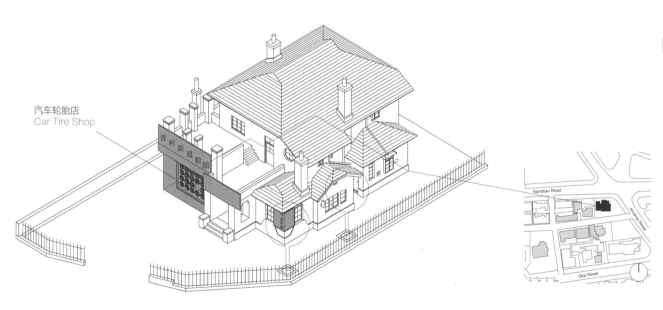

汽车轮胎店
Car Tire Shop

开口部改造
Door and Window

围墙改造
Fence Renovation

墙体粉刷
Painting

阳台改造
Balcony Renovation

连接加建
Connective Addition

独立加建
Unaided Addition

绿化
Green

已拆除
Demolished

保存良好
Well-preserved

军产住宅
Military Property

改变用途
Conversion

住户增加
Residents Plus

A

Right elevation

Front elevation

13	地址 Address	建成年代 Built in	保护等级 Protection Level	层数 Floors	类型 Type	用途 Use	结构 Structure	长×宽×高（m） L×W×H
	林风街 8 号 No.8 Linfeng Street	1922 前 Before 1922	一级 Level 1	2 2	独栋住宅 Detached House	居住 Living	砖木结构 Masonry-timber	17.2×10.9×7.3 17.2×10.9×7.3

现有 1 户 3 口之家居住，1983 年入住。建筑内部格局，一层有 5 个房间，分别为老人用房、卧室以及三个厅室；东侧的一个厅室内带有壁柜，有一部楼梯可通向二层的卧室。木质地板显得陈旧，排水设备相对简陋。可能由于年久失修的原因，外墙皮部分脱落，墙身的某些部位出现了纵向裂痕。

There is a household of three people living here from 1983. The interior pattern of the building is as below: there are five rooms on the first floor, which are a room for the elderly, one bedroom and three living room; one of the living room on the east has a closet inside; there is a staircase to the bedrooms on the second floor. The wooden floors look timeworn, and the drainage facility is relatively simple and crude. Due to the lack of renovation, part of the outer wall face peeled off. There are some longitudinal cracks on the surface of part of the walls.

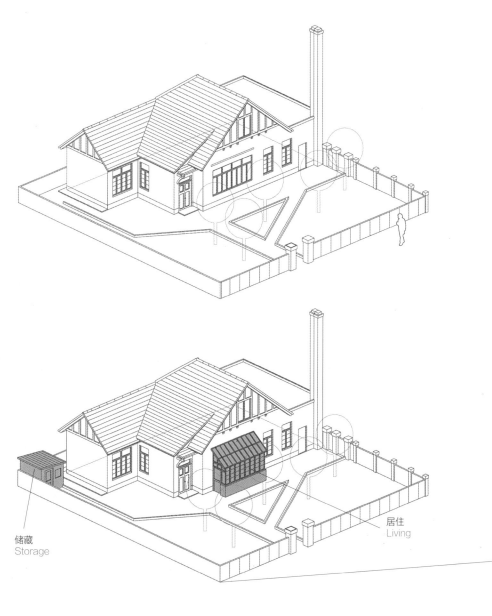

储藏
Storage

居住
Living

开口部改造 Door and Window	围墙改造 Fence Renovation	墙体粉刷 Painting	阳台改造 Balcony Renovation	连接加建 Connective Addition	独立加建 Unaided Addition	绿化 Green	已拆除 Demolished	保存良好 Well-preserved	军产住宅 Military Property	改变用途 Conversion	住户增加 Residents Plus	

ARMY

A

	地址 Address	建成年代 Built in	保护等级 Protection Level	层数 Floors	类型 Type	用途 Use	结构 Structure	长×宽×高（m） L×W×H
14	南山街 35 号 No.35 Nanshan Street	1940 前 Before 1940	二级 Level 2	2 2	集合住宅 Collective Housing	居住 Living	砖木结构 Masonry-timber	18.0×12.0×8.8 18.0×12.0×8.8

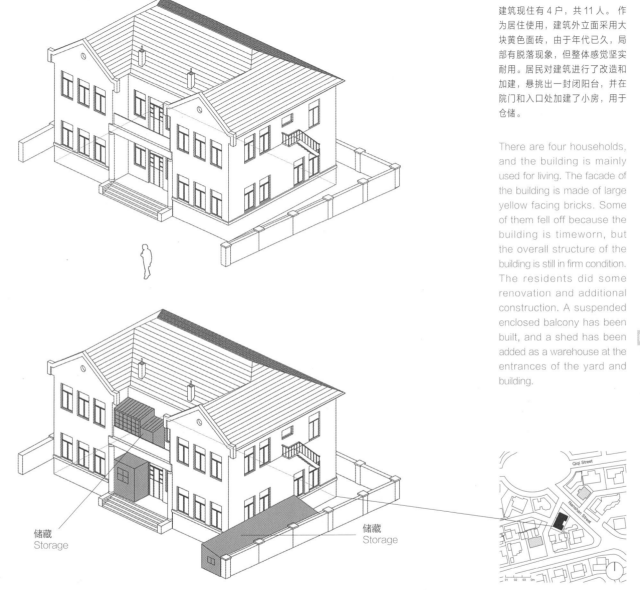

建筑现住有 4 户，共 11 人。作为居住使用，建筑外立面采用大块黄色面砖，由于年代已久，局部有脱落现象，但整体感觉坚实耐用。居民对建筑进行了改造和加建，悬挑出一封闭阳台，并在院门和入口处加建了小房，用于仓储。

There are four households, and the building is mainly used for living. The facade of the building is made of large yellow facing bricks. Some of them fell off because the building is timeworn, but the overall structure of the building is still in firm condition. The residents did some renovation and additional construction. A suspended enclosed balcony has been built, and a shed has been added as a warehouse at the entrances of the yard and building.

041

储藏
Storage

储藏
Storage

| 开口部改造 | 围墙改造 | 墙体粉刷 | 阳台改造 | 连接加建 | 独立加建 | 绿化 | 已拆除 | 保存良好 | 军产住宅 | 改变用途 | 住户增加 |
| Door and Window | Fence Renovation | Painting | Balcony Renovation | Connective Addition | Unaided Addition | Green | Demolished | Well-preserved | Military Property | Conversion | Residents Plus |

A

Left elevation

Back elevation

Detail

Detail

	地址 Address	建成年代 Built in	保护等级 Protection Level	层数 Floors	类型 Type	用途 Use	结构 Structure	长×宽×高（m） L×W×H
15	南山路杏林街交汇 Nanshan Road and Xinglin Street	1940 前 Before 1940	一级 Level 1	3 /–1 3 /–1	独栋住宅 Detached House	居住 Living	砖木结构 Masonry-timber	16.0×15.8×14 16.0×15.8×14

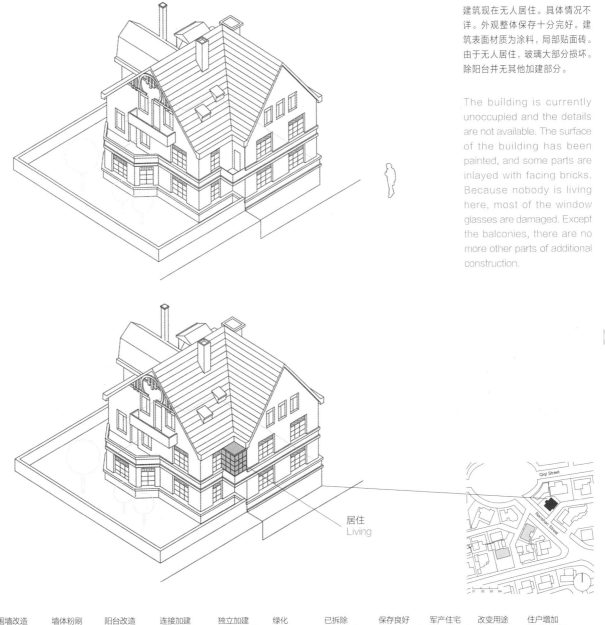

建筑现在无人居住。具体情况不详。外观整体保存十分完好。建筑表面材质为涂料，局部贴面砖。由于无人居住，玻璃大部分损坏。除阳台并无其他加建部分。

The building is currently unoccupied and the details are not available. The surface of the building has been painted, and some parts are inlayed with facing bricks. Because nobody is living here, most of the window glasses are damaged. Except the balconies, there are no more other parts of additional construction.

居住
Living

Qiqi Street

Nanshan Street

开口部改造
Door and Window

围墙改造
Fence Renovation

墙体粉刷
Painting

阳台改造
Balcony Renovation

连接加建
Connective Addition

独立加建
Unaided Addition

绿化
Green

已拆除
Demolished

保存良好
Well-preserved

军产住宅
Military Property

改变用途
Conversion

住户增加
Residents Plus

A

Left elevation

Right elevation

Back elevation

Front elevation

	地址 Address	建成年代 Built in	保护等级 Protection Level	层数 Floors	类型 Type	用途 Use	结构 Structure	长×宽×高（m） L×W×H
16	青林街 9 号 No.9 Qinglin Street	1922 前 Before 1922	二级 Level 2	2 2	独栋住宅 Detached House	居住 Living	砖木结构 Masonry-timber	15.3×9.7×9.9 15.3×9.7×9.9

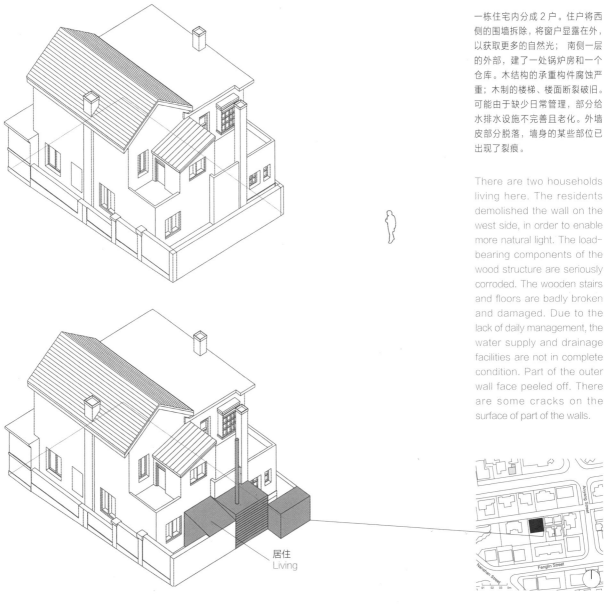

一栋住宅内分成 2 户。住户将西侧的围墙拆除，将窗户显露在外，以获取更多的自然光；南侧一层的外部，建了一处锅炉房和一个仓库。木结构的承重构件腐蚀严重；木制的楼梯、楼面断裂破旧。可能由于缺少日常管理，部分给水排水设施不完善且老化。外墙皮部分脱落，墙身的某些部位已出现了裂痕。

There are two households living here. The residents demolished the wall on the west side, in order to enable more natural light. The load-bearing components of the wood structure are seriously corroded. The wooden stairs and floors are badly broken and damaged. Due to the lack of daily management, the water supply and drainage facilities are not in complete condition. Part of the outer wall face peeled off. There are some cracks on the surface of part of the walls.

居住
Living

开口部改造
Door and Window

围墙改造
Fence Renovation

墙体粉刷
Painting

阳台改造
Balcony Renovation

连接加建
Connective Addition

独立加建
Unaided Addition

绿化
Green

已拆除
Demolished

保存良好
Well-preserved

军产住宅
Military Property

改变用途
Conversion

住户增加
Residents Plus

A

Left elevation

Detail

Detail

Back elevat

	地址 Address	建成年代 Built in	保护等级 Protection Level	层数 Floors	类型 Type	用途 Use	结构 Structure	长×宽×高（m） L×W×H
17	安阳街 80-84 号 No.80-84 Anyang Street	1940 前 Before 1940	三级 Level 3	2 2	集合住宅 Collective Housing	居住 Living	砖混结构 Masonry-concrete	20.0×14.5×9.8 20.0×14.5×9.8

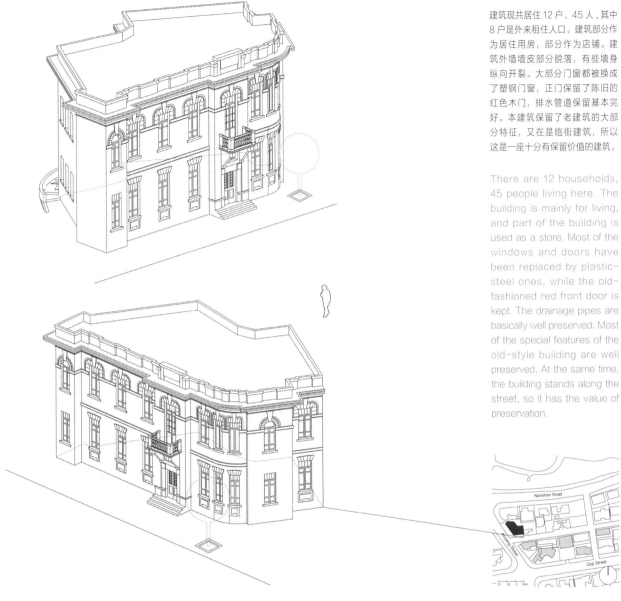

建筑现共居住 12 户、45 人，其中 8 户是外来租住人口。建筑部分作为居住用房，部分作为店铺。建筑外墙墙皮部分脱落，有些墙身纵向开裂。大部分门窗都被换成了塑钢门窗，正门保留了陈旧的红色木门，排水管道保留基本完好。本建筑保留了老建筑的大部分特征，又是临街建筑，所以这是一座十分有保留价值的建筑。

There are 12 households, 45 people living here. The building is mainly for living, and part of the building is used as a store. Most of the windows and doors have been replaced by plastic-steel ones, while the old-fashioned red front door is kept. The drainage pipes are basically well preserved. Most of the special features of the old-style building are well preserved. At the same time, the building stands along the street, so it has the value of preservation.

047

开口部改造	围墙改造	墙体粉刷	阳台改造	连接加建	独立加建	绿化	已拆除	保存良好	军产住宅	改变用途	住户增加
Door and Window	Fence Renovation	Painting	Balcony Renovation	Connective Addition	Unaided Addition	Green	Demolished	Well-preserved	Military Property	Conversion	Residents Plus

B

建筑部分作为居住用房，部分作为店铺。住宅内现共居住6户、14人，其中5户是外来租住人口。建筑内部结构已遭破坏，设施陈旧，上下水设施不完善，一些排水管道从中间折断，木制楼梯楼板也已断裂破旧。外墙墙皮脱落，有些墙身纵向开裂。

Part of the building is used for living and the other part is used for stores. There are 6 households, 14 people in total now living in this building, five of which are nonlocal tenants. The interior structure of the building has been badly damaged. The facilities are timeworn and the water supply and drainage facilities are not complete. At the same time, some of the drainage pipes are broke in the middle, and the wooden stairs as well as the floors are damaged. Parts of the exterior wall has flaked off, and there are some longitudinal cracks on the surface of part of the walls.

居住
Living

店铺
Shop

居住
Living

048

	地址 Address	建成年代 Built in	保护等级 Protection Level	层数 Floors	类型 Type	用途 Use	结构 Structure	长×宽×高（m） L×W×H
18	安阳街 88 号 No.88 Anyang Street	1922-1945 1922-1945	三级 Level 3	2 2	集合住宅 Collective Housing	居住 / 店铺 Living / Shop	砖木结构 Masonry-timber	17.5×7.2×6.9 17.5×7.2×6.9

Left elevation

Right elevation

Detail

开口部改造
Door and Window

围墙改造
Fence Renovation

墙体粉刷
Painting

阳台改造
Balcony Renovation

连接加建
Connective Addition

独立加建
Unaided Addition

绿化
Green

已拆除
Demolished

保存良好
Well-preserved

军产住宅
Military Property

改变用途
Conversion

住户增加
Residents Plus

B

建筑形制等级较高，细节处理精美，原为德国领事馆要员居住，中华人民共和国成立后为解放军部队后勤部长居住。现已被 3 户居民使用，一层 2 户，二层 1 户。建筑东侧原有一口井，是过去主人用于浇灌自己的花园，后期由于土地使用的需要被填埋。

The building has a high-level appearance and exquisite details. It was originally inhabited by the members of the German Consulate. There was a well on the east side of the building which was used by the owner to water his garden in the past. In the later stage, due to the need of land, it has been buried.

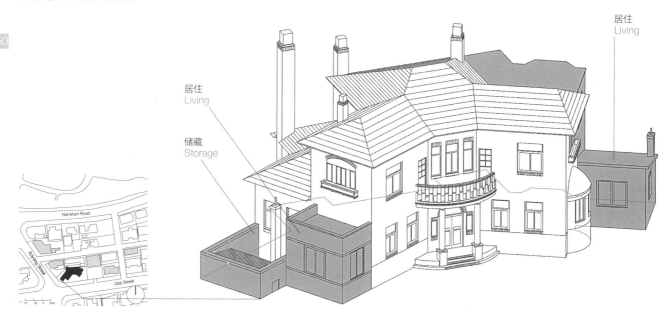

居住
Living

居住
Living

储藏
Storage

Nanshan Road

Anning Street

Qiqi Street

	地址 Address	建成年代 Built in	保护等级 Protection Level	层数 Floors	类型 Type	用途 Use	结构 Structure	长×宽×高（m） L×W×H
19	七七街80号 No.80 Qiqi Street	1922 前 Before 1922	一级 Level 1	2 2	独栋住宅 Detached House	居住 Living	砖木结构 Masonry-timber	16.6×13.3×9.6 16.6×13.3×9.6

开口部改造	围墙改造	墙体粉刷	阳台改造	连接加建	独立加建	绿化	已拆除	保存良好	军产住宅	改变用途	住户增加
Door and Window	Fence Renovation	Painting	Balcony Renovation	Connective Addition	Unaided Addition	Green	Demolished	Well-preserved	Military Property	Conversion	Residents Plus

B

Detail

Detail

Detail

Detail

Elevation

20	地址 Address	建成年代 Built in	保护等级 Protection Level	层数 Floors	类型 Type	用途 Use	结构 Structure	长 × 宽 × 高（m） L × W × H
	七七街 82 号 No.82 Qiqi Street	1922-1945 1922-1945	三级 Level 3	3 3	集合住宅 Collective Housing	居住 Living	砖木结构 Masonry-timber	15.0×12.8×12.6 15.0×12.8×12.6

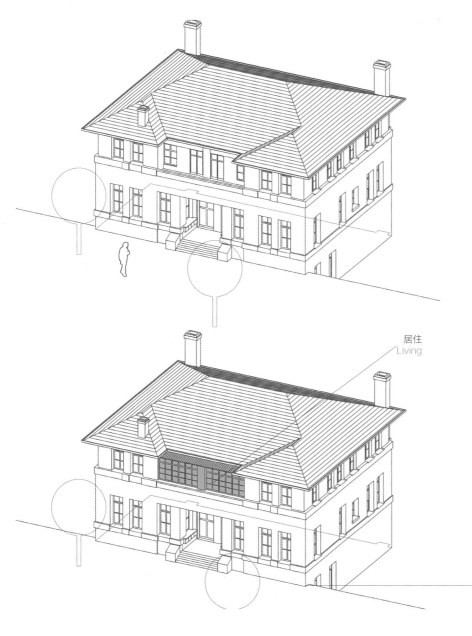

原居住 2 户，现居住 8 户。地势南高北低，南侧入口直通二层。由于住户逐年增多，内部空间分隔已经非常凌乱，但依然保持着内廊贯通的空间布局。建筑外立面除部分门窗外几乎未作改动，仅在三层南向将阳台封闭，改造为卧室。

There used to be two, but now eight households living here. The terrain here is high in the south and low in the north. There is an entrance on the west directly connecting with the second floor. The interior compartmentation is in disorder for the increasing number of residents, but the spatial layout of the through corridor is still well kept. There are few change of the exterior facade except some doors and windows. Only the balcony to the south on the third floor was changed to a bedroom.

053

居住
Living

| 开口部改造 Door and Window | 围墙改造 Fence Renovation | 墙体粉刷 Painting | 阳台改造 Balcony Renovation | 连接加建 Connective Addition | 独立加建 Unaided Addition | 绿化 Green | 已拆除 Demolished | 保存良好 Well-preserved | 军产住宅 Military Property | 改变用途 Conversion | 住户增加 Residents Plus |

B

Left elevation

Back elevation

Right elevation

Elevation

21	地址 Address	建成年代 Built in	保护等级 Protection Level	层数 Floors	类型 Type	用途 Use	结构 Structure	长×宽×高（m） L×W×H
	七七街 86 号 No.86 Qiqi Street	1922-1945 1922-1945	三级 Level 3	2 2	集合住宅 Collective Housing	居住 Living	砖木结构 Masonry-timber	16.0×5.8×6.1 16.0×5.8×6.1

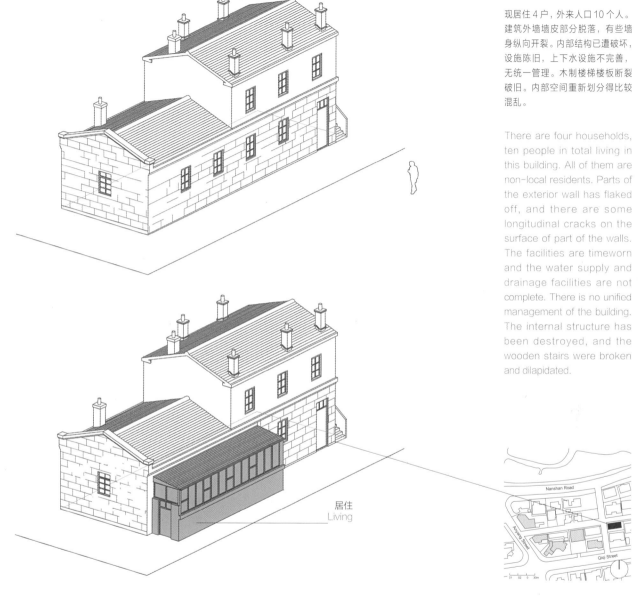

现居住 4 户，外来人口 10 个人。建筑外墙墙皮部分脱落，有些墙身纵向开裂。内部结构已遭破坏，设施陈旧，上下水设施不完善，无统一管理。木制楼梯楼板断裂破旧。内部空间重新划分得比较混乱。

There are four households, ten people in total living in this building. All of them are non-local residents. Parts of the exterior wall has flaked off, and there are some longitudinal cracks on the surface of part of the walls. The facilities are timeworn and the water supply and drainage facilities are not complete. There is no unified management of the building. The internal structure has been destroyed, and the wooden stairs were broken and dilapidated.

055

居住
Living

开口部改造	围墙改造	墙体粉刷	阳台改造	连接加建	独立加建	绿化	已拆除	保存良好	军产住宅	改变用途	住户增加
Door and Window	Fence Renovation	Painting	Balcony Renovation	Connective Addition	Unaided Addition	Green	Demolished	Well-preserved	Military Property	Conversion	Residents Plus

B

Detail

Back Elevation

Detail

Detail

	地址 Address	建成年代 Built in	保护等级 Protection Level	层数 Floors	类型 Type	用途 Use	结构 Structure	长×宽×高（m） L×W×H
22	青林街 35 号 No.35 Qinglin Street	1922-1945 1922-1945	三级 Level 3	3 3	独栋住宅 Detached House	居住 Living	砖木结构 Masonry-timber	17.0×10.7×10. 17.0×10.7×10.

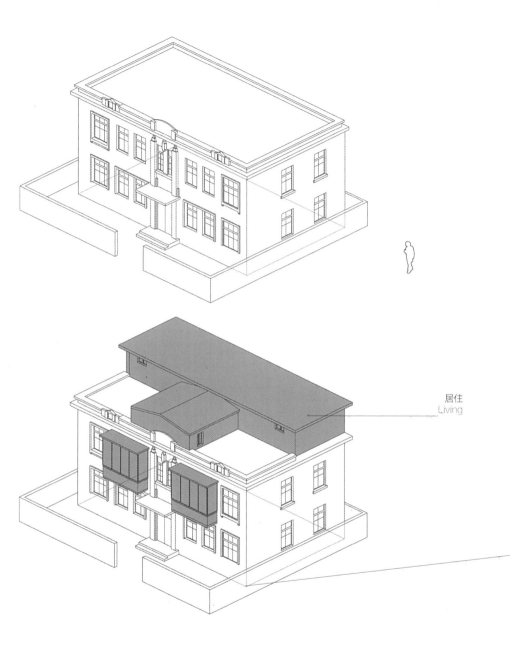

原为苏联官员高级住宅寓所，现一栋住宅内住8户。为提高空间使用率，将原阁楼空间增加为2户居住。外墙材料主要为砖和混凝土抹灰，虽然历时久远，但依然保持完好。空间格局基本保持原始状态，单元户型较大，楼梯间宽敞，内置有取暖设施。住户自发在梯道内养花，改善环境。

This building was used as high-class apartments for a Soviet officer, and now there are eight households living here. In order to improve space utilization, the previous loft has been renovated for two households to live. The exterior walls were made with a combination of bricks and concrete. Though having a long history, the building is still in good condition. The residents grow flowers in the stairway to improve the environment.

居住
Living

开口部改造
Door and Window

围墙改造
Fence Renovation

墙体粉刷
Painting

阳台改造
Balcony Renovation

连接加建
Connective Addition

独立加建
Unaided Addition

绿化
Green

已拆除
Demolished

保存良好
Well-preserved

军产住宅
Military Property

改变用途
Conversion

住户增加
Residents Plus

B

Detail

Detail

Back elevation

	地址 Address	建成年代 Built in	保护等级 Protection Level	层数 Floors	类型 Type	用途 Use	结构 Structure	长 × 宽 × 高（m） L × W × H
23	青林街凡尔赛会馆 Versalles Club	1922-1945 1922-1945	一级 Level 1	2 2	公建 Public Building	娱乐 / 居住 Recreation / Living	砖混结构 Masonry-timber	40.0 × 26.6 ×18. 40.0 × 26.6 ×18.

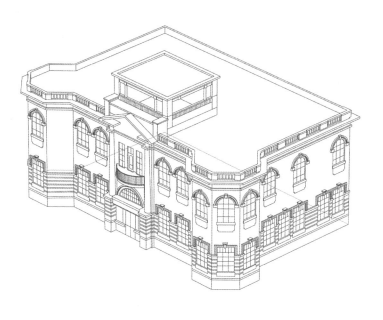

建筑原来的使用性质为办公楼，现为私人会馆。外表面为繁复的欧式线脚，立面造型左右对称，建筑基本保持了原始的风貌。目前该建筑处于闲置状态。

This building was originally used as an office block. Now it is a private club. The outer surface of the building is decorated with delicate European-style patterns. The facades are symmetrical on both sides. The building has been maintained with its original style and features. The building is unoccupied at the moment.

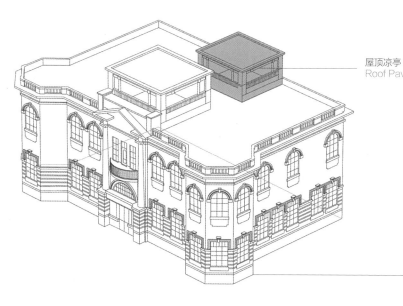

屋顶凉亭
Roof Pavilion

开口部改造	围墙改造	墙体粉刷	阳台改造	连接加建	独立加建	绿化	已拆除	保存良好	军产住宅	改变用途	住户增加
Door and Window	Fence Renovation	Painting	Balcony Renovation	Connective Addition	Unaided Addition	Green	Demolished	Well-preserved	Military Property	Conversion	Residents Plus

B

Left elevation

Back elevation

Right elevation

Detail

24	地址 Address	建成年代 Built in	保护等级 Protection Level	层数 Floors	类型 Type	用途 Use	结构 Structure	长 × 宽 × 高（m） L × W × H
	南山路 207 号 No.207 Nanshan Road	1922-1945 1922-1945	三级 Level 3	2 2	独栋住宅 Detached House	居住 Living	砖木结构 Masonry-timber	12.2 × 12.0 × 8.8 12.2 × 12.0 × 8.8

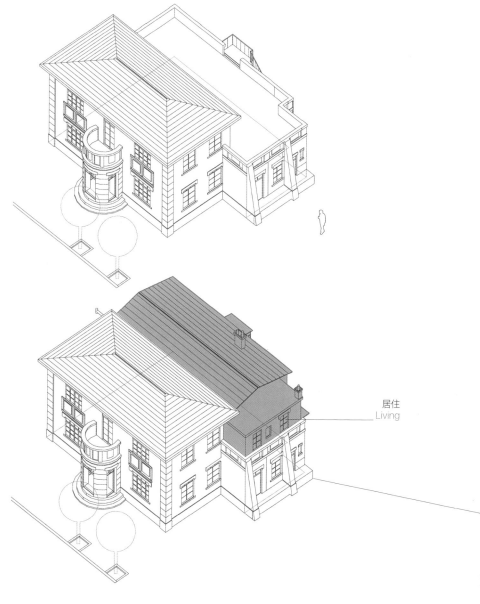

原居7户、20人。测绘时，此
建筑正处于拆迁状态中，有些门
窗已被拆卸。仅1户人家居住。
建筑内部空间划分比较混乱，设
施陈旧。由于建筑外墙皮部分脱
落，临街部分进行抹灰处理。

There used to be seven
households, twenty people
living here. During mapping,
this building was already
under demolition. Some of
the doors and windows has
been disassembled. At the
moment, only one household
is still living in the building. The
interior compartmentation
is in disorder, and most of
the facilities are timeworn.
Because most of the outer
wall face of the building
peeled off, the street-facing
side of the building has been
plastered.

居住
Living

| 开口部改造
Door and Window | 围墙改造
Fence Renovation | 墙体粉刷
Painting | 阳台改造
Balcony Renovation | 连接加建
Connective Addition | 独立加建
Unaided Addition | 绿化
Green | 已拆除
Demolished | 保存良好
Well-preserved | 军产住宅
Military Property | 改变用途
Conversion | 住户增加
Residents Plus |

   +   + +   B

C 区老建筑
Site C Old Buildings

Left elevation

Detail

Right elevation

	地址 Address	建成年代 Built in	保护等级 Protection Level	层数 Floors	类型 Type	用途 Use	结构 Structure	长×宽×高（m） L×W×H
25	南山路 209 号 No.209 Nanshan Road	1940 前 Before 1940	三级 Level 3	2 2	独栋住宅 Detached House	居住 / 店铺 Living / Shop	砖木结构 Masonry-timber	16.4×14.4×7.0 16.4×14.4×7.0

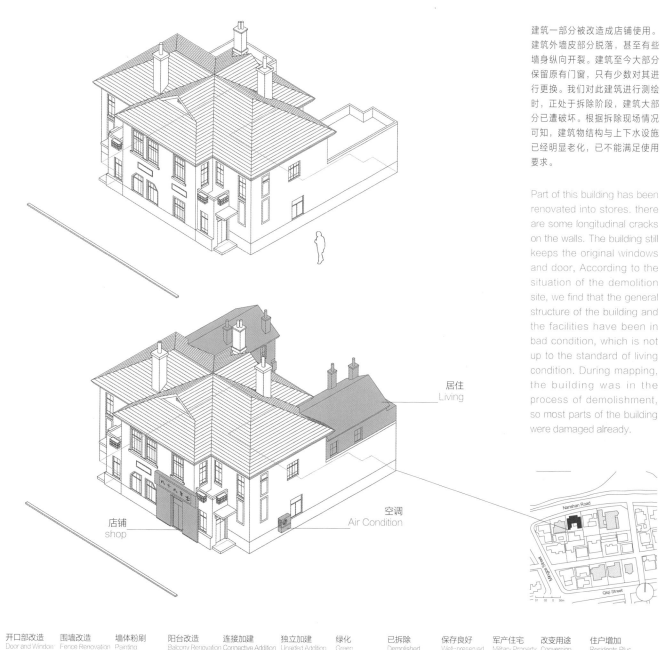

建筑一部分被改造成店铺使用。建筑外墙皮部分脱落，甚至有些墙身纵向开裂。建筑至今大部分保留原有门窗，只有少数对其进行更换。我们对此建筑进行测绘时，正处于拆除阶段，建筑大部分已遭破坏。根据拆除现场情况可知，建筑物结构与上下水设施已经明显老化，已不能满足使用要求。

Part of this building has been renovated into stores. there are some longitudinal cracks on the walls. The building still keeps the original windows and door, According to the situation of the demolition site, we find that the general structure of the building and the facilities have been in bad condition, which is not up to the standard of living condition. During mapping, the building was in the process of demolishment, so most parts of the building were damaged already.

063

居住
Living

空调
Air Condition

店铺
shop

| 开口部改造 | 围墙改造 | 墙体粉刷 | 阳台改造 | 连接加建 | 独立加建 | 绿化 | 已拆除 | 保存良好 | 军产住宅 | 改变用途 | 住户增加 |
| Door and Window | Fence Renovation | Painting | Balcony Renovation | Connective Addition | Unaided Addition | Green | Demolished | Well-preserved | Military Property | Conversion | Residents Plus |

C

原建筑为医院，内部重新分隔后，作为住宅使用，共有18户居民。建筑整体保存完好，墙体、屋顶等无明显的破坏迹象。建筑由于功能上的改变，空间上无法很好满足居住功能。建筑内部分构造已老化，上下水、采暖等问题严重。此建筑具有较高的保护价值。

This building was used as a hospital before. Later on, the interior space was rearranged for dwelling, and now there are 18 households living here. It is generally well preserved, and there are no obvious damages of the walls and roof. Due to the changes of the building's function, the spaces can't meet the living function very well. The interior structure of the building is timeworn. However, the building has high preserving value.

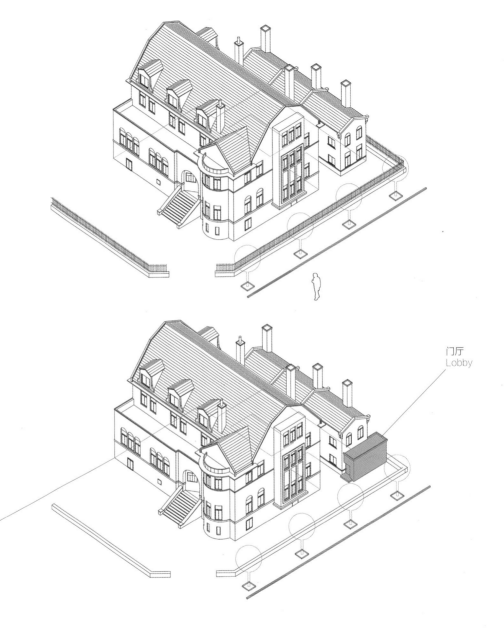

门厅
Lobby

	地址 Address	建成年代 Built in	保护等级 Protection Level	层数 Floors	类型 Type	用途 Use	结构 Structure	长×宽×高（m） L×W×H
26	南山路 215 号 No.215 Nanshan Road	1934-1938 1934-1938	三级 Level 3	4 4	集合住宅 Collective Housing	居住 Living	砖木结构 Masonry-timber	21.9×23.8×15 21.9×23.8×15

Left elevation

Detail

Interior

Detail

| 开口部改造 Door and Window | 围墙改造 Fence Renovation | 墙体粉刷 Painting | 阳台改造 Balcony Renovation | 连接加建 Connective Addition | 独立加建 Unaided Addition | 绿化 Green | 已拆除 Demolished | 保存良好 Well-preserved | 军产住宅 Military Property | 改变用途 Conversion | 住户增加 Residents Plus |

C

此建筑原为苏联高级干部别墅，初建时配有完整的上下水设施和锅炉房，等级较高。现租住约5户人家。建筑内部空间被住户划分得比较混乱，外墙皮部分脱落，但结构完好。

This building used to be the villa for a soviet senior officer. Originally, it was equipped with complete water supply and drainage facilities as well as a boiler room, which was a high-leveled building at that time. Now five households are living here, and all of them are tenants. The interior space of the building has been divided by the residents in disorder. Part of the exterior wall has flaked off, but the general structure is well preserved.

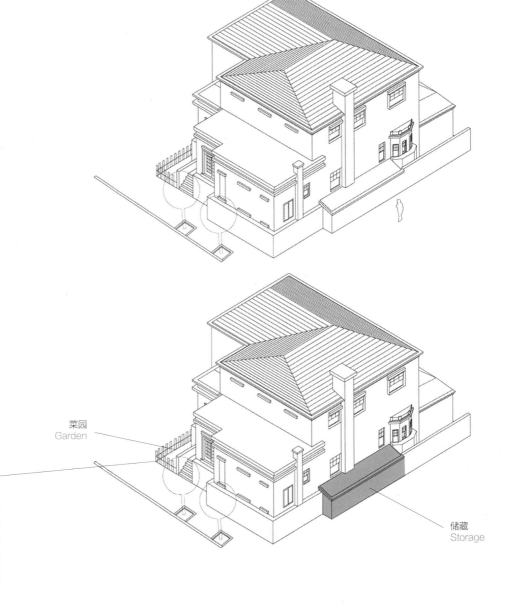

菜园
Garden

储藏
Storage

Nanshan Road
Wuwu Road
Qiqi Street

	地址 Address	建成年代 Built in	保护等级 Protection Level	层数 Floors	类型 Type	用途 Use	结构 Structure	长×宽×高（m） L×W×H
27	南山路217号 No.217 Nanshan Road	1922-1945 1922-1945	三级 Level 3	2 2	独栋住宅 Detached House	居住 Living	砖木结构 Masonry-timber	17.2×12.6×9.2 17.2×12.6×9.2

Detail

Elevation

Elevation

Elevation

开口部改造	围墙改造	墙体粉刷	阳台改造	连接加建	独立加建	绿化	已拆除	保存良好	军产住宅	改变用途	住户增加
Door and Window	Fence Renovation	Painting	Balcony Renovation	Connective Addition	Unaided Addition	Green	Demolished	Well-preserved	Military Property	Conversion	Residents Plus

C

Left elevation

Detail

Detail

Detail

Detail

28	地址 Address	建成年代 Built in	保护等级 Protection Level	层数 Floors	类型 Type	用途 Use	结构 Structure	长×宽×高（m） L×W×H
	南山路 221/223 号 No.221/223 Nanshan Road	1922–1945 1922–1945	三级 Level 3	2 2	独栋住宅 Detached House	居住 Living	砖木结构 Masonry-timber	18.4×9.7×7.2 18.4×9.7×7.2

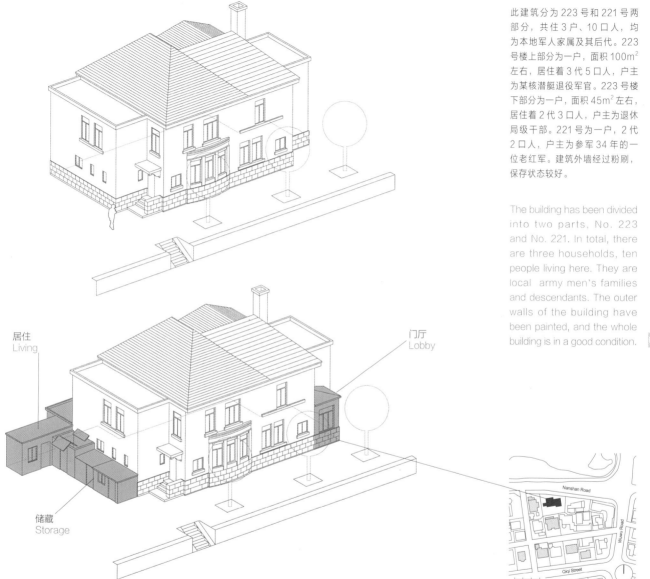

此建筑分为 223 号和 221 号两部分，共住 3 户、10 口人，均为本地军人家属及其后代。223 号楼上部分为一户，面积 100m² 左右，居住着 3 代 5 口人，户主为某核潜艇退役军官。223 号楼下部分为一户，面积 45m² 左右，居住着 2 代 3 口人，户主为退休局级干部。221 号为一户，2 代 2 口人，户主为参军 34 年的一位老红军。建筑外墙经过粉刷，保存状态较好。

The building has been divided into two parts, No. 223 and No. 221. In total, there are three households, ten people living here. They are local army men's families and descendants. The outer walls of the building have been painted, and the whole building is in a good condition.

069

居住
Living

储藏
Storage

门厅
Lobby

| 开口部改造 Door and Window | 围墙改造 Fence Renovation | 墙体粉刷 Painting | 阳台改造 Balcony Renovation | 连接加建 Connective Addition | 独立加建 Unaided Addition | 绿化 Green | 已拆除 Demolished | 保存良好 Well-preserved | 军产住宅 Military Property | 改变用途 Conversion | 住户增加 Residents Plus |

建筑现为水电煤气缴费网点，并
且作为拆迁指挥部使用。原有木
质楼梯保存较好，上下水设备齐
全，环境舒适。建筑整体保存状
态良好，室内木作做工精良。据
传，最早为日本一高级官员居所。

The building is now used as
a payment network for water,
electricity and gas, and as
the demolition headquarter at
the same time. The original
wooden stares are well
preserved. In general, the
building offers a comfortable
living condition. The overall
structure of the building is well
preserved, and the interior
woodwork of the building is
made in high quality.

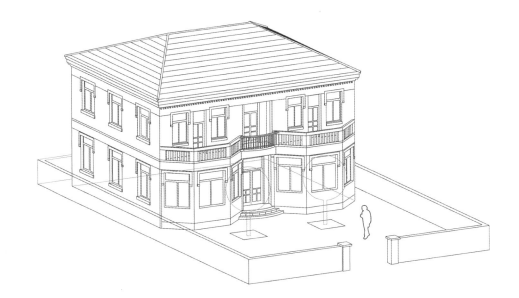

29	地址 Address	建成年代 Built in	保护等级 Protection Level	层数 Floors	类型 Type	用途 Use	结构 Structure	长×宽×高（m） L×W×H
	哈尔滨街76号 No.76 Haerbin Street	1937 1937	二级 Level 2	2 2	独栋住宅 Detached House	银行 Bank	砖木结构 Masonry-timber	14.0×15.0×8.6 14.0×15.0×8.6

| 开口部改造
Door and Window | 围墙改造
Fence Renovation | 墙体粉刷
Painting | 阳台改造
Balcony Renovation | 连接加建
Connective Addition | 独立加建
Unaided Addition | 绿化
Green | 已拆除
Demolished | 保存良好
Well-preserved | 军产住宅
Military Property | 改变用途
Conversion | 住户增加
Residents Plus |

   + + + +   **C**

Left elevation

Detail

Right elevation

Back elevation

30	地址 Address	建成年代 Built in	保护等级 Protection Level	层数 Floors	类型 Type	用途 Use	结构 Structure	长 × 宽 × 高（m） L × W × H
	七七街 98-1 号 No.98-1 Qiqi Street	1922-1945 1922-1945	一级 Level 1	3 3	独栋住宅 Detached House	居住 Living	砖木结构 Masonry-timber	16.0 × 13.0 × 7.4 16.0 × 13.0 × 7.4

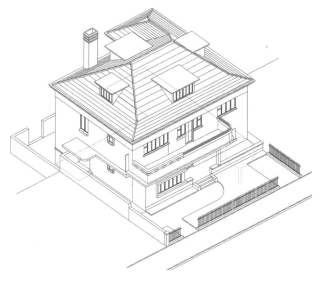

现居住有两户军人家庭。由于地形南高北低，因此南侧为 2 层，北侧局部 3 层。北侧和南侧各加建有不足 10m² 的储藏室。门窗也基本没有大的改动，仍为木质平开窗。内部空间被划分为两部分，为两家人共同居住。

There are two military families living here. Because the terrain is high in the south and low in the north, there are two floors in the south part of the building while three floors in the north. Two storage rooms of less than 10 square meters each have been built on both sides. The doors are basically unchanged, as well as the windows, which are still wooden casement style. The interior space is divided into two parts for two families to live together.

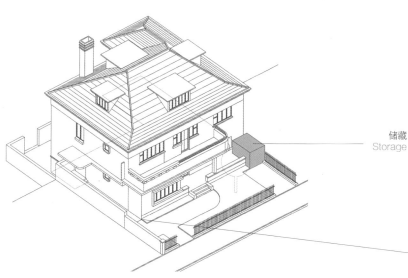

储藏
Storage

开口部改造
Door and Window

围墙改造
Fence Renovation

墙体粉刷
Painting

阳台改造
Balcony Renovation

连接加建
Connective Addition

独立加建
Unaided Addition

绿化
Green

已拆除
Demolished

保存良好
Well-preserved

军产住宅
Military Property

改变用途
Conversion

住户增加
Residents Plus

C

原为日本高官居住用房，中华人民共和国成立后为区政府官员住房，现在为普通居民房，内有3户、12人。上下水完好，且有暖气，内部管道铺设于地板下，净化了室内空间。

It used to be a residence for senior officials in Japan. Now it is a common residence with 3 households and 12 people. The water supply and drainage facilities are in good condition, and the heating system has been equipped. The internal pipes are laid under the floor, which purifies the indoor space.

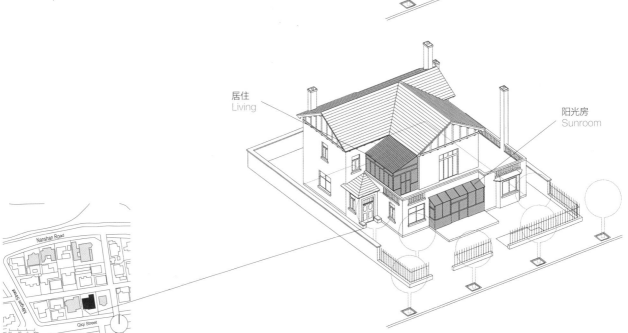

居住
Living

阳光房
Sunroom

Nanshan Road

Qiqi Street

	31	地址 Address	建成年代 Built in	保护等级 Protection Level	层数 Floors	类型 Type	用途 Use	结构 Structure	长×宽×高（m） L×W×H
		七七街100号 No.100 Qiqi'Street	1922-1945 1922-1945	二级 Level 2	2 2	独栋住宅 Detached House	居住 Living	砖木结构 Masonry-timber	17.5×17.0×9.7 17.5×17.0×9.7

Back elevation

Detail

Detail

Left elevation

原为当年满铁株式会社社长的寓所，中华人民共和国成立后普通民众居住于此。此建筑地上2层，地下1层。地上经过改造分成4户，供10人居住，并且每户有各自的厨卫。地下为锅炉用房，面积达100多平方米，负责此建筑的供水和供暖。

This building used to be the home for the president of Manchuria Railway Co., Ltd. After founding of People's Republic of China, the building was converted to residence for common people. The building has two floors above ground and one floor underground. The two upper floors have been reconstructed into four dwellings of ten people in total. All the dwellings have their respective kitchens and toilets. The underground floor is used as a 100m² boiler room for heating and hot water supply.

居住 / 储藏
Living / Storage

Nanshan Road
Minze Street
Qiqi Street

	32	地址 Address	建成年代 Built in	保护等级 Protection Level	层数 Floors	类型 Type	用途 Use	结构 Structure	长×宽×高（m） L×W×H
		七七街102号 No.102 Qiqi Street	1922-1945 1922-1945	二级 Level 2	3 3	独栋住宅 Detached House	居住 Living	砖木结构 Masonry-timber	9.8×8.9×9.8 9.8×8.9×9.8

Elevation

Right elevation

Detail

Detail

Detail

| 开口部改造
Door and Window | 围墙改造
Fence Renovation | 墙体粉刷
Painting | 阳台改造
Balcony Renovation | 连接加建
Connective Addition | 独立加建
Unaided Addition | 绿化
Green | 已拆除
Demolished | 保存良好
Well-preserved | 军产住宅
Military Property | 改变用途
Conversion | 住户增加
Residents Plus |

B

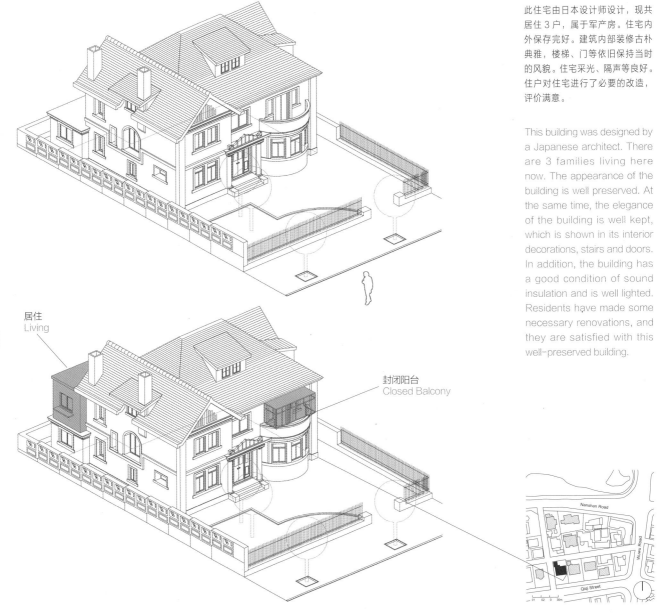

此住宅由日本设计师设计，现共居住 3 户，属于军产房。住宅内外保存完好。建筑内部装修古朴典雅，楼梯、门等依旧保持当时的风貌。住宅采光、隔声等良好。住户对住宅进行了必要的改造，评价满意。

This building was designed by a Japanese architect. There are 3 families living here now. The appearance of the building is well preserved. At the same time, the elegance of the building is well kept, which is shown in its interior decorations, stairs and doors. In addition, the building has a good condition of sound insulation and is well lighted. Residents have made some necessary renovations, and they are satisfied with this well-preserved building.

居住
Living

封闭阳台
Closed Balcony

Nanshan Road

Wuwu Road

Qiqi Street

	地址 Address	建成年代 Built in	保护等级 Protection Level	层数 Floors	类型 Type	用途 Use	结构 Structure	长×宽×高（m） L×W×H
33	七七街 104 号 No.104 Qiqi Street	1922-1945 1922-1945	一级 Level 1	2 2	独栋住宅 Detached House	居住 Living	砖木结构 Masonry-timber	13.6×13.7×10 13.6×13.7×10

Detail

Interior detail

Left elevation

开口部改造 Door and Window	围墙改造 Fence Renovation	墙体粉刷 Painting	阳台改造 Balcony Renovation	连接加建 Connective Addition	独立加建 Unaided Addition	绿化 Green	已拆除 Demolished	保存良好 Well-preserved	军产住宅 Military Property	改变用途 Conversion	住户增加 Residents Plus

C

Detail

Detail

Detail

Detail

34	地址 Address	建成年代 Built in	保护等级 Protection Level	层数 Floors	类型 Type	用途 Use	结构 Structure	长×宽×高（m） L×W×H
	七七街108号 No.108 Qiqi Street	1922-1945 1922-1945	一级 Level 1	2 /-1 2 /-1	独栋住宅 Detached House	居住 Living	砖木结构 Masonry-timber	15.0×12.0×7.8 15.0×12.0×7.

此建筑现在处于空置状态，内外部装修已结束，今后用途不详。室内石膏板顶棚，地面铺地毯，木楼梯等细部精良，做工讲究，还保持原有韵味。

This building is unoccupied now, and its future use is unknown, but the interior and exterior decorations have been completed, The building is applied with a gypsum board ceiling, and the floor is covered with carpet. The wooden stairs and other details were made with exquisite workmanship, which maintains the original charm of the building.

食堂
Dining Room

开口部改造	围墙改造	墙体粉刷	阳台改造	连接加建	独立加建	绿化	已拆除	保存良好	军产住宅	改变用途	住户增加	
Door and Window	Fence Renovation	Painting	Balcony Renovation	Connective Addition	Unaided Addition	Green	Demolished	Well-preserved	Military Property	Conversion	Residents Plus	C

Left elevation

Detail

Detail

Detail

	地址 Address	建成年代 Built in	保护等级 Protection Level	层数 Floors	类型 Type	用途 Use	结构 Structure	长×宽×高（m） L×W×H
35	七七街110号 No.110 Qiqi Street	1940 前 Before 1940	三级 Level 3	2 2	独栋住宅 Detached House	居住 Living	砖木结构 Masonry-timber	12.3×8.8×9.5 12.3×8.8×9.5

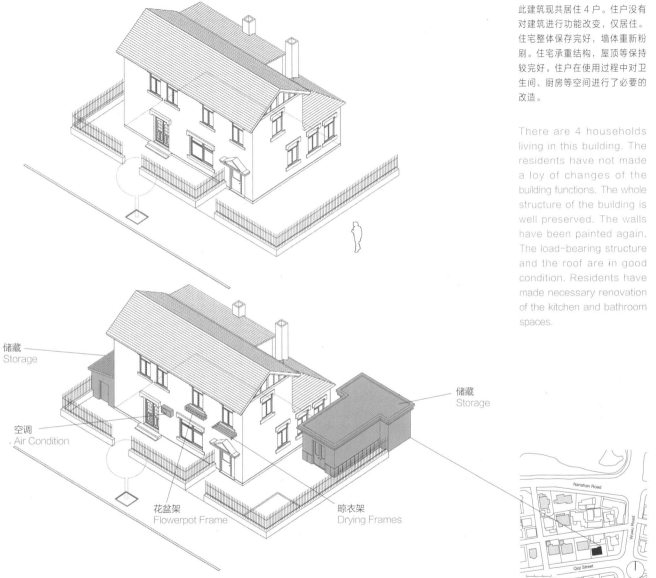

此建筑现共居住 4 户。住户没有对建筑进行功能改变，仅居住。住宅整体保存完好，墙体重新粉刷。住宅承重结构，屋顶等保持较完好。住户在使用过程中对卫生间、厨房等空间进行了必要的改造。

There are 4 households living in this building. The residents have not made a loy of changes of the building functions. The whole structure of the building is well preserved. The walls have been painted again. The load-bearing structure and the roof are in good condition. Residents have made necessary renovation of the kitchen and bathroom spaces.

083

储藏
Storage

储藏
Storage

空调
Air Condition

花盆架
Flowerpot Frame

晾衣架
Drying Frames

开口部改造 Door and Window	围墙改造 Fence Renovation	墙体粉刷 Painting	阳台改造 Balcony Renovation	连接加建 Connective Addition	独立加建 Unaided Addition	绿化 Green	已拆除 Demolished	保存良好 Well-preserved	军产住宅 Military Property	改变用途 Conversion	住户增加 Residents Plus	C

	地址 Address	建成年代 Built in	保护等级 Protection Level	层数 Floors	类型 Type	用途 Use	结构 Structure	长×宽×高（m） L×W×H
36	明泽街 88 号 No.88 Mingze Street	1922-1945 1922-1945	三级 Level 3	2 2	独栋住宅 Detached House	居住 Living	砖木结构 Masonry-timber	17.9×10.1×8.3 17.9×10.1×8.3

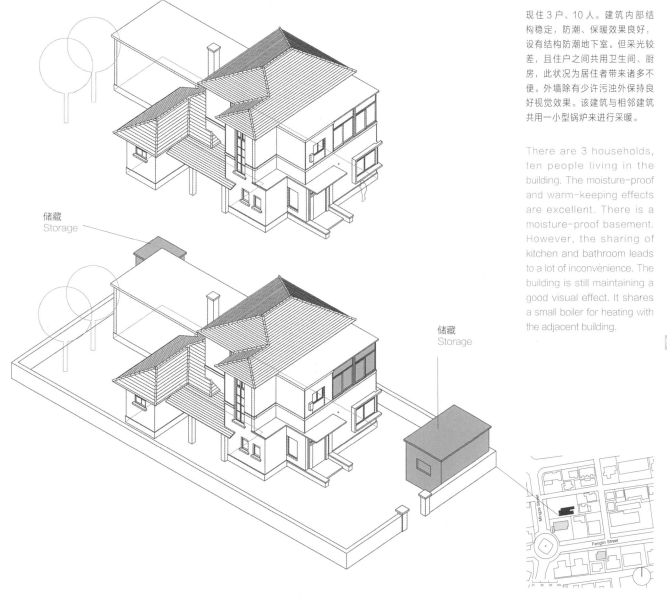

现住 3 户、10 人。建筑内部结构稳定，防潮、保暖效果良好，设有结构防潮地下室。但采光较差，且住户之间共用卫生间、厨房，此状况为居住者带来诸多不便。外墙除有少许污浊外保持良好视觉效果。该建筑与相邻建筑共用一小型锅炉来进行采暖。

There are 3 households, ten people living in the building. The moisture-proof and warm-keeping effects are excellent. There is a moisture-proof basement. However, the sharing of kitchen and bathroom leads to a lot of inconvenience. The building is still maintaining a good visual effect. It shares a small boiler for heating with the adjacent building.

储藏
Storage

储藏
Storage

开口部改造
Door and Window

围墙改造
Fence Renovation

墙体粉刷
Painting

阳台改造
Balcony Renovation

连接加建
Connective Addition

独立加建
Unaided Addition

绿化
Green

已拆除
Demolished

保存良好
Well-preserved

军产住宅
Military Property

改变用途
Conversion

住户增加
Residents Plus

C

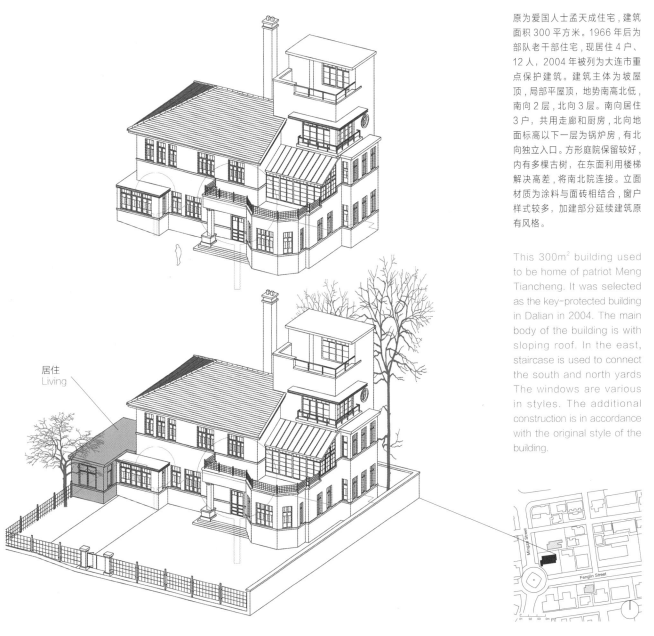

原为爱国人士孟天成住宅，建筑面积 300 平方米。1966 年后为部队老干部住宅，现居住 4 户、12 人，2004 年被列为大连市重点保护建筑。建筑主体为坡屋顶，局部平屋顶，地势南高北低，南向 2 层，北向 3 层。南向居住 3 户，共用走廊和厨房，北向地面标高以下一层为锅炉房，有北向独立入口。方形庭院保留较好，内有多棵古树，在东面利用楼梯解决高差，将南北院连接。立面材质为涂料与面砖相结合，窗户样式较多，加建部分延续建筑原有风格。

This 300m² building used to be home of patriot Meng Tiancheng. It was selected as the key-protected building in Dalian in 2004. The main body of the building is with sloping roof. In the east, staircase is used to connect the south and north yards The windows are various in styles. The additional construction is in accordance with the original style of the building.

居住
Living

	地址 Address	建成年代 Built in	保护等级 Protection Level	层数 Floors	类型 Type	用途 Use	结构 Structure	长×宽×高（m） L×W×H
37	枫林街 36 号 No.36 Fenglin Street	1922-1945 1922-1945	一级 Level 1	3 3	独栋住宅 Detached House	居住 Living	砖木结构 Masonry-timber	22.1×12.8×15. 22.1×12.8×15.

 开口部改造
Door and Window

 围墙改造
Fence Renovation

墙体粉刷
Painting

阳台改造
Balcony Renovation

连接加建
Connective Addition

独立加建
Unaided Addition

绿化
Green

已拆除
Demolished

保存良好
Well-preserved

军产住宅
Military Property

改变用途
Conversion

住户增加
Residents Plus

Detail

Detail

Left elevation

Back elevation

C

Left elevation

Detail

Right elevation

Detail

38	地址 Address	建成年代 Built in	保护等级 Protection Level	层数 Floors	类型 Type	用途 Use	结构 Structure	长 × 宽 × 高（m） L × W × H
	五五路 97 号 No.97 Wuwu Road	1922 前 Before 1922	三级 Level 3	2 2	独栋住宅 Detached House	居住 / 店铺 Living / Shop	砖木结构 Masonry-timber	27.8×15.5×9.8 27.8×15.5×9.8

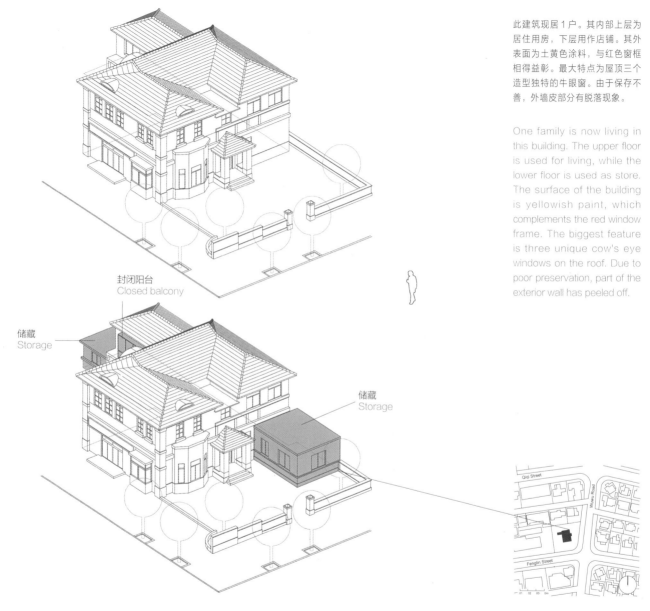

此建筑现居 1 户。其内部上层为居住用房，下层用作店铺。其外表面为土黄色涂料，与红色窗框相得益彰。最大特点为屋顶三个造型独特的牛眼窗。由于保存不善，外墙皮部分有脱落现象。

One family is now living in this building. The upper floor is used for living, while the lower floor is used as store. The surface of the building is yellowish paint, which complements the red window frame. The biggest feature is three unique cow's eye windows on the roof. Due to poor preservation, part of the exterior wall has peeled off.

089

封闭阳台
Closed balcony

储藏
Storage

储藏
Storage

C

现居 1 户两位老人，户主原为部队高级军官。由于维护得当，建筑内部设施保持良好的使用状态。个别窗户有透风现象。外部有一处连接加建，作储藏用。外墙装饰有杏黄色涂料，整体色泽良好，并无脱落裂缝现象。

There are two elders living in the building, The building is in good condition though one or two windows are poor in air tightness. Outside of the building there is a connected additional construction, which is used for storage. The outer walls of the building are painted in beautiful almond yellow. The whole building has a bright color, with no traces of cracks or peel-offs.

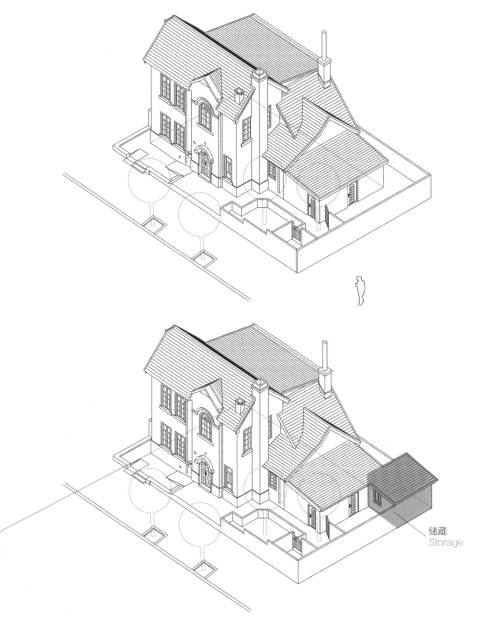

储藏
Storage

	地址 Address	建成年代 Built in	保护等级 Protection Level	层数 Floors	类型 Type	用途 Use	结构 Structure	长 × 宽 × 高 (m) L × W × H
39	枫林街 37 号 No.37 Fenglin Street	1922-1945 1922-1945	二级 Level 2	2 2	独栋住宅 Detached House	居住 Living	砖木结构 Masonry-timber	19.8 × 13.2 × 9.8 19.8 × 13.2 × 9.8

Front elevation

Detail

Right elevation

Left elevation

Detail

开口部改造 Door and Window	围墙改造 Fence Renovation	墙体粉刷 Painting	阳台改造 Balcony Renovation	连接加建 Connective Addition	独立加建 Unaided Addition	绿化 Green	已拆除 Demolished	保存良好 Well-preserved	军产住宅 Military Property	改变用途 Conversion	住户增加 Residents Plus	

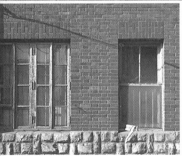

	地址 Address	建成年代 Built in	保护等级 Protection Level	层数 Floors	类型 Type	用途 Use	结构 Structure	长×宽×高（m） L×W×H
40	望海街 36 号 No.36 Wanghai Street	1922-1945 1922-1945	二级 Level 2	2 2	独栋住宅 Detached House	居住 Living	砖木结构 Masonry-timber	12.5×11.2×10.8 12.5×11.2×10.8

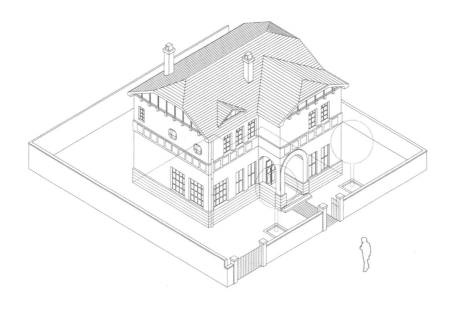

此住宅产权归部队所有，除西侧增加入户楼梯之外，基本都保持了建筑原貌。但因为年久失修，屋顶部分保温性能不佳，冬天比较寒冷。住宅内部空间因住户增加而被划分为几间小间。院子西北侧建有一小型锅炉房，供住户采暖使用。

The property of this building belongs to the army. The appearance is barely changed except an additional stair in the west. Due to the lack of maintenance, the roof has a poor heat preservation performance during winter. The interior space has been divided into several rooms because of the increasing number of the residents. There is a small boiler room in the northwest yard for the heating supply.

093

储藏
Storage

空调
Air Condition

储藏
Storage

楼梯
Stair

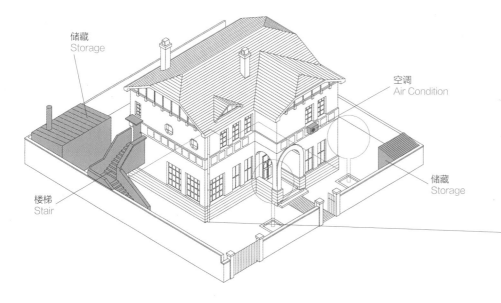

开口部改造	围墙改造	墙体粉刷	阳台改造	连接加建	独立加建	绿化	已拆除	保存良好	军产住宅	改变用途	住户增加
Door and Window	Fence Renovation	Painting	Balcony Renovation	Connective Addition	Unaided Addition	Green	Demolished	Well-preserved	Military Property	Conversion	Residents Plus

C

Left elevation

Detail

Front elevation

Detail

41	地址 Address	建成年代 Built in	保护等级 Protection Level	层数 Floors	类型 Type	用途 Use	结构 Structure	长×宽×高（m） L×W×H
	南山路 230 号 No.230 Nanshan Road	1927 前 Before 1927	三级 Level 3	2 2	独栋住宅 Detached House	居住 Living	砖木结构 Masonry-timber	14.2×11.0×8.0 14.2×11.0×8.0

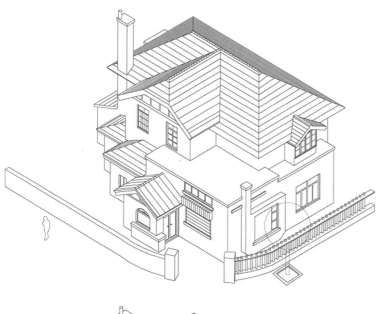

此建筑原为日本人居住的独户别墅，住户身份不详。现共租住 10 余户，内部划分成紧密的小间，布局混乱，光线很暗，门窗做过局部维护。建筑外墙皮陈旧，久未更新。

This building once was a Japanese single-family villa, now are rented to more than 10 households. The interior is divided into compact little rooms, which are all dimly-lit. The layout is quite confusing. The doors and windows have been partially maintained. The outer wall of the building is old and has not been restored for a long time.

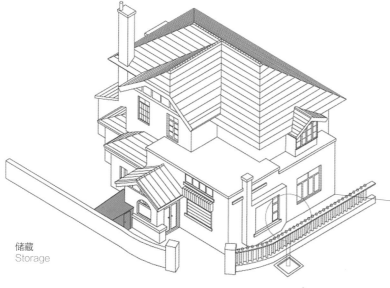

储藏
Storage

开口部改造	围墙改造	墙体粉刷	阳台改造	连接加建	独立加建	绿化	已拆除	保存良好	军产住宅	改变用途	住户增加
Door and Window	Fence Renovation	Painting	Balcony Renovation	Connective Addition	Unaided Addition	Green	Demolished	Well-preserved	Military Property	Conversion	Residents Plus

Left elevation

Detail

Front elevation

Detail

42	地址 Address	建成年代 Built in	保护等级 Protection Level	层数 Floors	类型 Type	用途 Use	结构 Structure	长 × 宽 × 高（m） L×W×H
	南山路 233 号 No.233 Nanshan Road	1922 前 Before 1922	三级 Level 3	2 2	独栋住宅 Detached House	居住 / 店铺 Living / Shop	砖木结构 Masonry-timber	19.6×12.5×6.8 19.6×12.5×6.8

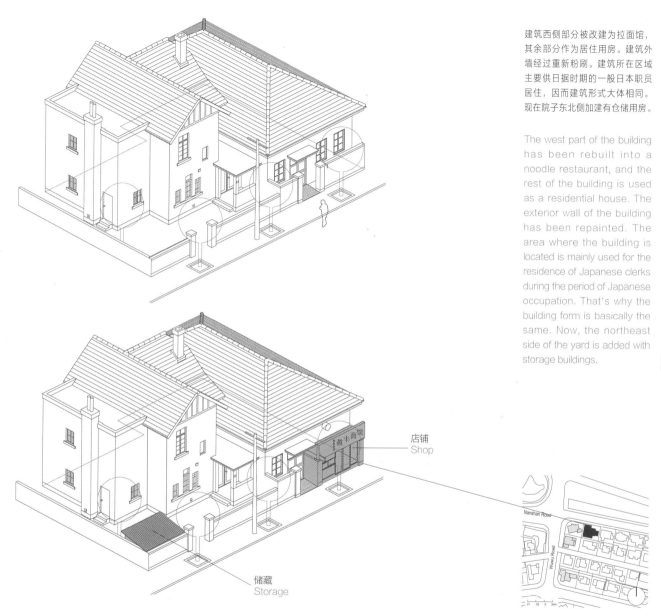

建筑西侧部分被改建为拉面馆，其余部分作为居住用房。建筑外墙经过重新粉刷。建筑所在区域主要供日据时期的一般日本职员居住，因而建筑形式大体相同。现在院子东北侧加建有仓储用房。

The west part of the building has been rebuilt into a noodle restaurant, and the rest of the building is used as a residential house. The exterior wall of the building has been repainted. The area where the building is located is mainly used for the residence of Japanese clerks during the period of Japanese occupation. That's why the building form is basically the same. Now, the northeast side of the yard is added with storage buildings.

097

店铺
Shop

储藏
Storage

Nanshan Road

Wuwu Road

开口部改造
Door and Window

围墙改造
Fence Renovation

墙体粉刷
Painting

阳台改造
Balcony Renovation

连接加建
Connective Addition

独立加建
Unaided Addition

绿化
Green

已拆除
Demolished

保存良好
Well-preserved

军产住宅
Military Property

改变用途
Conversion

住户增加
Residents Plus

D

住宅一小部分改成店铺，大部分仍用于居住。住宅外墙面无破损、开裂等情况。建筑的承重结构、屋顶等保持良好。住宅内部上下水管道已老化。据住户介绍，根据需要对卫生间、厨房等空间进行了改造。目前此栋住宅已拆除。

A small part of the house has been converted into store, while most part of the house are still used for living. There are no damages or cracks on the exterior wall of the house. The load-bearing structure of the building are kept in good condition. The upper and lower water pipes inside the house are time-worn. According to the introduction of the residents, the toilet, kitchen and other spaces should be transformed according to the needs. At present, the house has been demolished.

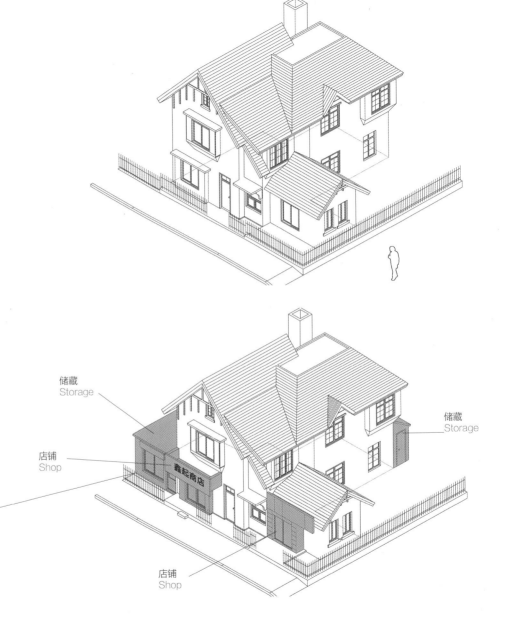

储藏
Storage

储藏
Storage

店铺
Shop

鑫起商店

店铺
Shop

Nanshan Road

Wuwu Road

	地址 Address	建成年代 Built in	保护等级 Protection Level	层数 Floors	类型 Type	用途 Use	结构 Structure	长×宽×高（m） L×W×H
43	南山路 201 号 No.201 Nanshan Road	1940 前 Before 1940	二级 Level 2	2 2	独栋住宅 Detached House	居住 / 店铺 Living / Shop	砖木结构 Masonry-timber	11.5×10.8×9.0 11.5×10.8×9.0

Detail

Detail

Detail

Detail

开口部改造
Door and Window

围墙改造
Fence Renovation

墙体粉刷
Painting

阳台改造
Balcony Renovation

连接加建
Connective Addition

独立加建
Unaided Addition

绿化
Green

已拆除
Demolished

保存良好
Well-preserved

军产住宅
Military Property

改变用途
Conversion

住户增加
Residents Plus

 +   + + +   D

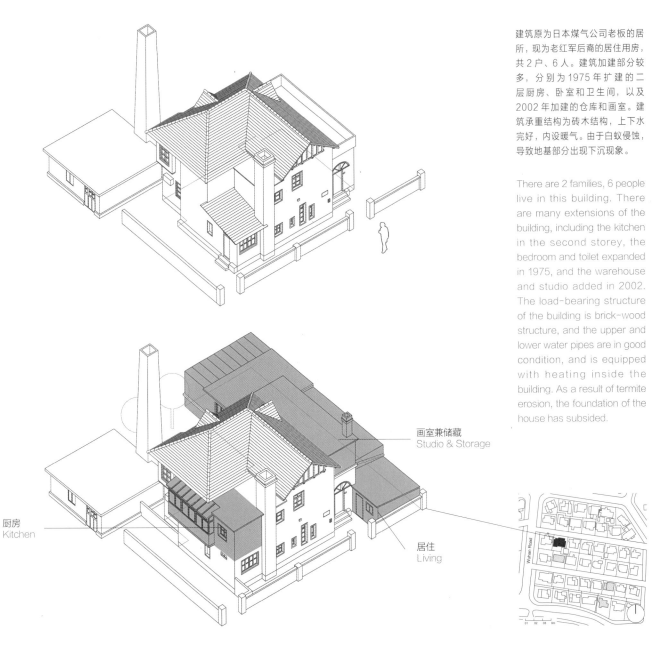

建筑原为日本煤气公司老板的居所，现为老红军后裔的居住用房，共2户、6人。建筑加建部分较多，分别为1975年扩建的二层厨房、卧室和卫生间，以及2002年加建的仓库和画室。建筑承重结构为砖木结构，上下水完好，内设暖气。由于白蚁侵蚀，导致地基部分出现下沉现象。

There are 2 families, 6 people live in this building. There are many extensions of the building, including the kitchen in the second storey, the bedroom and toilet expanded in 1975, and the warehouse and studio added in 2002. The load-bearing structure of the building is brick-wood structure, and the upper and lower water pipes are in good condition, and is equipped with heating inside the building. As a result of termite erosion, the foundation of the house has subsided.

画室兼储藏
Studio & Storage

厨房
Kitchen

居住
Living

	地址 Address	建成年代 Built in	保护等级 Protection Level	层数 Floors	类型 Type	用途 Use	结构 Structure	长 × 宽 × 高（m） L × W × H
44	高阳路 41 号 No.41 Gaoyang Road	1922 前 Before 1922	三级 Level 3	2 2	独栋住宅 Detached House	居住 Living	砖木结构 Masonry-timber	16.0 × 9.6 × 6.9 16.0 × 9.6 × 6.9

Interior

Detail

Back elevation

Detail

| 开口部改造
Door and Window | 围墙改造
Fence Renovation | 墙体粉刷
Painting | 阳台改造
Balcony Renovation | 连接加建
Connective Addition | 独立加建
Unaided Addition | 绿化
Green | 已拆除
Demolished | 保存良好
Well-preserved | 军产住宅
Military Property | 改变用途
Conversion | 住户增加
Residents Plus | D |

住宅内现居住2户、8人，均为外来租住人口。为了满足居住需要，整栋建筑面积增加了近30%。加建后建筑外墙经过粉刷，整体比较协调。内部结构破坏较少，锅炉房单独供暖，保温性能良好，但室内光线不足，通风较差，且有漏雨。

There are 2 families, 8 people living in this people, In order to meet the residential needs, another 30% of building area has been added. After the additional construction, the exterior wall of the building was painted, making the whole of the building more harmonious. The internal structure is less damaged. The boiler room is equipped with separate heating and the heat retaining property is good, but the indoor light condition is insufficient, the ventilation is poor, and there is leakage when it rains.

居住
Living

居住
Living

卧室和卫生间
Bedroom & Toilet

	45	地址 Address	建成年代 Built in	保护等级 Protection Level	层数 Floors	类型 Type	用途 Use	结构 Structure	长×宽×高（m） L×W×H
		林景街2号 No.2 Linjing Street	1922前 Before 1922	三级 Level 3	2 2	独栋住宅 Detached House	居住 Living	砖木结构 Masonry-timber	11.5×10.0×6.3 11.5×10.0×6.3

 Front elevation

 Detail

 Detail

 Detail

开口部改造 Door and Window	围墙改造 Fence Renovation	墙体粉刷 Painting	阳台改造 Balcony Renovation	连接加建 Connective Addition	独立加建 Unaided Addition	绿化 Green	已拆除 Demolished	保存良好 Well-preserved	军产住宅 Military Property	改变用途 Conversion	住户增加 Residents Plus	

             D

建筑现有 5 户、10 人居住，可能由于年久失修，外墙皮部分脱落，墙身某些部位出现纵向开裂。建筑的加建部分延伸至院墙外缘，把院子分割为东西两部分，使两边住户能够拥有自己的独立空间。楼内排水设施简陋，无统一管理，木制楼梯和楼板破旧，为日常生活带来不便。

Currently, 5 households, 10 people in all live in this building. Some parts of the outer wall peeled off, and some parts of the wall cracked longitudinally. The additional part of the building extends to the outer edge of the courtyard wall, dividing the courtyard into east and west parts, so that the residents on both sides can have their own independent space. The drainage facilities in the building are simple and crude, and without unified management. The wooden stairs and floors are worn-out which brings inconvenience to daily life.

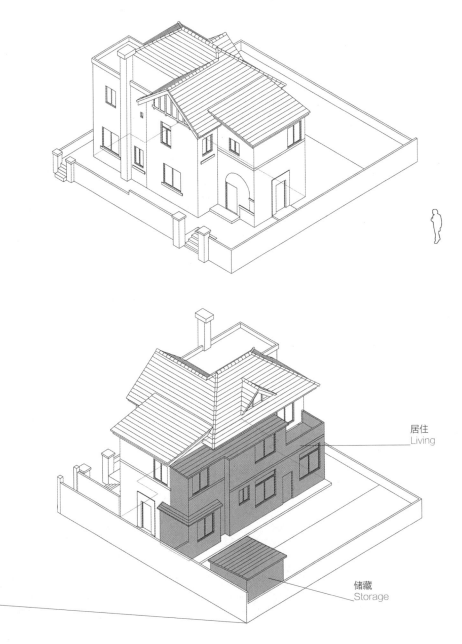

居住
Living

储藏
Storage

46	地址 Address	建成年代 Built in	保护等级 Protection Level	层数 Floors	类型 Type	用途 Use	结构 Structure	长×宽×高（m） L×W×H
	林景街 9 号 No.9 Linjing Road	1922 前 Before 1922	三级 Level 3	2 2	独栋住宅 Detached House	居住 Living	砖木结构 Masonry-timber	12.1×6.9×8.9 12.1×6.9×8.9

Detail

Front elevation

Detail

开口部改造	围墙改造	墙体粉刷	阳台改造	连接加建	独立加建	绿化	已拆除	保存良好	军产住宅	改变用途	住户增加
Door and Window	Fence Renovation	Painting	Balcony Renovation	Connective Addition	Unaided Addition	Green	Demolished	Well-preserved	Military Property	Conversion	Residents Plus

D

建筑现有 5 户、10 人居住。建筑外墙皮部分脱落，墙身结构破坏。部分给水排水设施不完善且老化。木制楼梯和楼板破旧，甚至有些部位出现断裂。

There are 5 households living in this building. Some parts of the outer wall peeled off, and the wall structure was damaged. Some of the water supply and drainage facilities are incomplete and worn-out. The wooden stairs and floors are dilapidated, and some parts have already cracked.

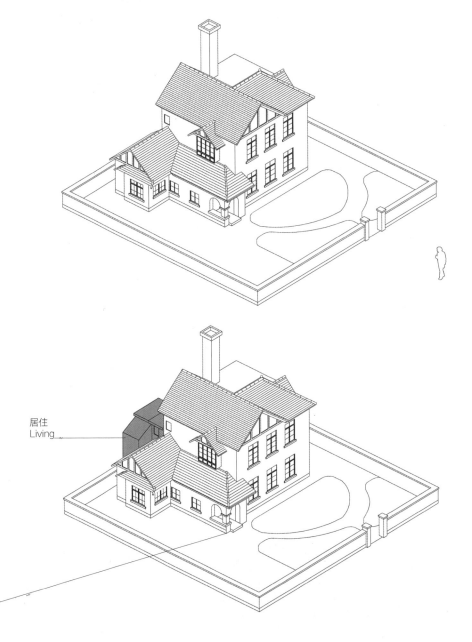

居住
Living

	地址 Address	建成年代 Built in	保护等级 Protection Level	层数 Floors	类型 Type	用途 Use	结构 Structure	长×宽×高（m） L×W×H
47	望海街 59 号 No.59 Wanghai Street	1922 前 Before 1922	三级 Level 3	2 2	独栋住宅 Detached House	居住 Living	砖木结构 Masonry-timber	13.6×10.3×7.8 13.6×10.3×7.8

| 开口部改造
Door and Window | 围墙改造
Fence Renovation | 墙体粉刷
Painting | 阳台改造
Balcony Renovation | 连接加建
Connective Addition | 独立加建
Unaided Addition | 绿化
Green | 已拆除
Demolished | 保存良好
Well-preserved | 军产住宅
Military Property | 改变用途
Conversion | 住户增加
Residents Plus |

D

老建筑属性

Attribute of the Old Buildings

分区	NO.	地址	建成年代	保护等级	产权	层数	类型	用途	结构	表面材质	长×宽×高（m）
A区	1	七七街76号	1922-1945	三级	单位	2	独栋住宅	居住	砖木结构	抹灰	10.7×8.0×9.9
	2	哈尔滨街45号	1922-1945	二级	个人	2	独栋住宅	居住	砖木结构	石材/抹灰	20.4×10.4×9.0
	3	哈尔滨街43号	1922-1945	二级	个人	3	集合住宅	居住	砖混结构	抹灰/清水砖	14.0×13.5×12.7
	4	哈尔滨街41号	1922-1945	三级	个人	1	独栋住宅	居住/店铺	砖木结构	石材/抹灰/清水砖	17.1×12×6.9
	5	南山街233号	1952	一级	个人	3	公建	幼儿园	砖木结构	清水砖墙	14.1×8.0×6.3
	6	七七街66号	1922-1945	一级	个人	2	独栋住宅	居住	砖木结构	清水砖/石材/抹灰	26.1×12.9×7.7
	7	哈尔滨街37号	1922-1945	二级	个人	2	独栋住宅	居住	砖木结构	抹灰	13.6×10.2×7.0
	8	南山路165号	1922-1945	二级	个人	2	集合住宅	居住/办公	砖木结构	抹灰	19.6×12.5×6.8
	9	哈尔滨街42/44/46/48号	1922-1945	三级	个人	2	集合住宅	居住	砖木结构	抹灰	21.2×9.0×8.5
	10	南山路181号	1922前	三级	个人		独栋住宅	居住	砖木结构	抹灰	13.2×7.8×6.8
	11	哈尔滨街50号	1922-1945	三级	个人		独栋住宅	居住	砖木结构	抹灰	11.4×11.4×4.8
	12	南山路182号	1922-1945	二级	个人	2	独栋住宅	居住/店铺	砖木结构	抹灰	17.6×13.4×9.8
	13	林风街8号	1922前	一级	个人	2	独栋住宅	居住	砖木结构	抹灰	17.2×10.9×7.3
	14	南山路35号	1940前	二级	军产	2	集合住宅	居住	砖木结构	面砖	18.0×12.0×8.8
	15	南山路与杏林街交汇	1940前	一级	个人	3/-1	独栋住宅	居住	砖木结构	面砖/抹灰	16.0×15.8×14.4
	16	青林街9号	1922前	二级	个人	2	独栋住宅	居住	砖木结构	抹灰	15.3×9.7×9.9
B区	17	安阳街80-84号	1940前	三级	个人	2	集合住宅	居住	砖混结构	抹灰	20×14.5×9.8
	18	安阳街88号	1922-1945	三级	个人	2	集合住宅	居住/店铺	砖木结构	抹灰	17.5×7.2×6.9
	19	七七街80号	1922前	一级	军产	2	独栋住宅	居住	砖木结构	抹灰	16.6×13.3×9.6
	20	七七街82号	1922-1945	三级	个人	3	集合住宅	居住	砖木结构	抹灰	15.0×12.8×12.6
	21	七七街86号	1922-1945	三级	个人	2	集合住宅	居住	砖木结构	抹灰	16.0×5.8×6.1
	22	青林街35号	1922-1945	三级	军产	3	独栋住宅	居住	砖木结构	清水砖墙/抹灰	17.0×10.7×10.7
	23	青林街凡尔赛会馆	1922-1945	一级	单位	2	公建	娱乐/居住	砖混结构	抹灰	40.0×26.6×18.2
	24	南山路207号	1922-1945	三级	个人	2	独栋住宅	居住	砖木结构	抹灰	12.2×12.0×8.8
C区	25	南山路209号	1940前	三级	个人	2	独栋住宅	居住/店铺	砖木结构	抹灰	16.4×14.4×7.0
	26	南山路215号	1934-1938	三级	个人	4	集合住宅	居住	砖木结构	抹灰/面砖	21.9×23.8×15.0
	27	南山路217号	1922-1945	三级	个人	2	独栋住宅	居住	砖木结构	抹灰	17.2×12.6×9.2
	28	南山路221/223号	1922-1945	三级	个人	2	独栋住宅	居住	砖木结构	抹灰	18.4×9.7×7.2
	29	哈尔滨街76号	1937	二级	军产	2	独栋住宅	银行	砖木结构	面砖/抹灰	14.0×15.0×8.6
	30	七七街98-1号	1922-1945	一级	个人	3	独栋住宅	居住	砖木结构	涂料/面砖	16.0×13.0×7.4
	31	七七街100号	1922-1945	二级	单位	2	独栋住宅	居住	砖木结构	抹灰	17.5×17.0×9.7
	32	七七街102号	1922-1945	二级	单位	3	独栋住宅	居住	砖木结构	抹灰	9.8×8.9×9.8
	33	七七街104号	1922-1945	一级	军产	2	独栋住宅	居住	砖木结构	抹灰	13.6×13.7×10.6
	34	七七街108号	1922-1945	一级	单位	2,-1	独栋住宅	居住	砖木结构	清水砖/抹灰	15.0×12.0×7.8
	35	七七街110号	1940前	三级	个人	2	独栋住宅	居住	砖木结构	抹灰	12.3×8.8×9.5
	36	明泽街88号	1922-1945	三级	军产	2	独栋住宅	居住	砖木结构	抹灰	17.9×10.1×8.3
	37	枫林街36号	1922-1945	一级	军产	3	独栋住宅	居住	砖木结构	面砖/抹灰	22.1×12.8×15.8
	38	五五路97号	1922前	三级	军产	2	独栋住宅	居住/店铺	砖木结构	抹灰	27.8×15.5×9.8
	39	枫林街37号	1922-1945	二级	军产	2	独栋住宅	居住	砖木结构	抹灰	19.8×13.2×9.8
	40	望海街36号	1922-1945	二级	军产	2	独栋住宅	居住	砖木结构	抹灰	12.5×11.2×10.8
	41	南山路230号	1927前	三级	个人	2	独栋住宅	居住	砖木结构	抹灰	14.2×11.0×8.0
D区	42	南山路233号	1922前	三级	个人	2	独栋住宅	居住/店铺	砖木结构	抹灰	19.6×12.5×6.8
	43	南山路201号	1940前	二级	个人	2	独栋住宅	居住/店铺	砖木结构	抹灰	11.5×10.8×9.0
	44	高阳街41号	1922前	三级	军产	2	独栋住宅	居住	砖木结构	抹灰	16.0×9.6×6.9
	45	林景街2号	1922前	三级	军产	2	独栋住宅	居住	砖木结构	抹灰	11.5×10.0×6.3
	46	林景街9号	1922前	三级	军产	2	独栋住宅	居住	砖木结构	抹灰	12.1×6.9×8.9
	47	望海街59号	1922前	三级	军产	2	独栋住宅	居住	砖木结构	抹灰	13.6×10.3×7.8

改造手法		构件更新				添加设施					已拆除		保存良好	军产	改变用途	住户增加
		建筑				建筑			环境							
加法	减法	门	窗户	围墙	粉刷	阳台	储藏	其他	储藏	绿化	2008年调查时	2011年调查时				
○		○	○	○	○		○						○			○
○			○				○	门罩	○						住宅-垃圾收购站（局部）	○
○			○				○						○		办公-住宅	
○			○		○		○	居住								○
○		○	○	○	○			雨篷		○			○			
○		○	○	○				办公			○				住宅-办公（局部）	
○			○					雨篷					○			○
○			○	○						○						○
○		○	○			○	○	居住/厨房								○
○				○			○	居住					○			○
○		○	○		○	○	○	居住		○						○
○			○	○						○					住宅-汽车轮胎店（局部）	○
○					○	○			○	○						○
○						○	○		○					○		○
○			○	○						○			○			○
													○		办公-住宅	
○		○	○		○	○		居住/入口							住宅-店铺（局部）	○
○	○		○		○		○	居住						○		○
○			○		○	○				○						○
○			○		○	○	○	居住								○
○			○		○	○	○	居住		○			○	○		
○		○	○		○			屋顶凉台		○					办公-会馆	
○					○			居住			○					○
○		○	○					居住			○				居住-店铺（局部）	○
○			○					门厅		○					医院-住宅	○
○			○	○			○			○						○
○			○		○		○	居住/门厅								○
			○									○	○	○	住宅—办公	
○		○	○				○			○						○
○		○	○	○	○	○		居住/阳光房		○						○
○			○	○		○	○	居住		○						○
○			○	○			○	封闭阳台		○			○	○		○
○		○	○	○	○	○	○	食堂		○			○	○	住宅-办公	
○			○	○	○		○		○	○			○			○
○			○			○	○		○	○			○	○		○
○			○	○			○			○			○	○		○
○		○	○		○	○	○	居住		○			○	○	住宅-店铺（局部）	
○			○	○			○			○			○	○		
○			○					楼梯	○	○			○	○		
○			○	○			○				○					○
○		○	○				○		○		○				住宅-店铺（局部）	
○		○	○	○	○		○				○				住宅-店铺（局部）	
○		○				○	○	居住/厨房/画室					○	○		○
○		○	○		○	○	○	居住	○				○	○		○
○		○	○			○	○			○				○		○
								居住		○				○		○

风格与构造
Style and Structure

风格与构造
Style and Structure

南山老街建筑多以日本传统住宅样式与欧美形式相融合的"近代折中主义"风格为主，创造了一种东西合璧的雅致外观，这种形式也符合东北寒冷地区砖墙为主的外观特点。在外观形式、结构构造、细部装饰等方面具有日据时期典型特征。

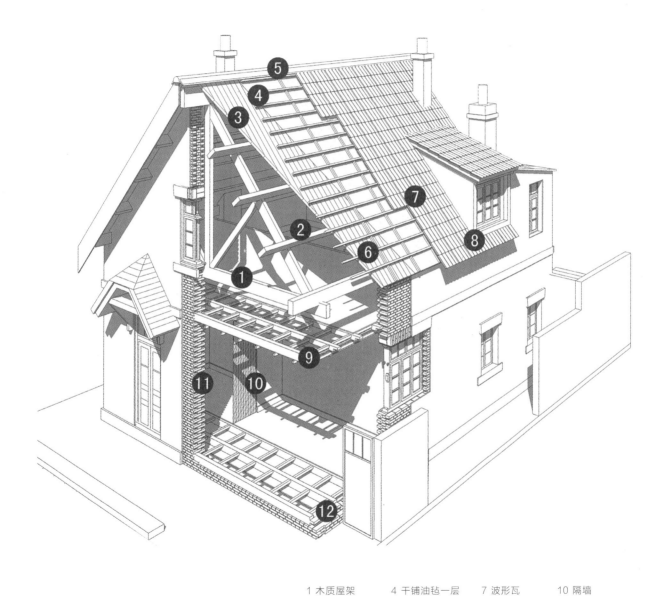

1 木质屋架　　4 干铺油毡一层　　7 波形瓦　　10 隔墙
2 方木檩条　　5 顺水条　　　　　8 老虎窗　　11 370 砖墙（外墙）
3 木格板　　　6 挂瓦条　　　　　9 木梁　　　12 防水层

结构与构造

主体结构：砖、石结构，以砖砌体结构为主，外墙砖厚 370mm。
内部构造：木结构为主，屋顶、楼梯、楼板均为木质构造，内隔墙多采用竹筋抹灰做法。

外观细部

建筑多采用折中主义手法，细部多样，形式丰富。外立面有老虎窗、外凸窗台或门廊等元素。门廊设计中，柱式风格有古埃及风格的柱头，也有古希腊陶立克风格等。
屋顶：四坡顶、人字顶、双折顶；
外墙：清水砖外墙、抹灰外墙、局部面砖装饰外墙；
门窗：木质门窗、木制百叶窗；
屋顶窗：老虎窗、三坡窗、三角楣窗、直窗等。

室外围栏与小品设计

住宅的院落设计中，围栏、室外踏步多以西式石材造型为主，手法上以现代风格处理，简洁、美观，比例尺度得体。

内部装饰

住宅内部装饰以木构做法与木制肌理为主，延续日本的起居特点。

典型细部
Typical Details

街区建筑

Buildings in the Blocks

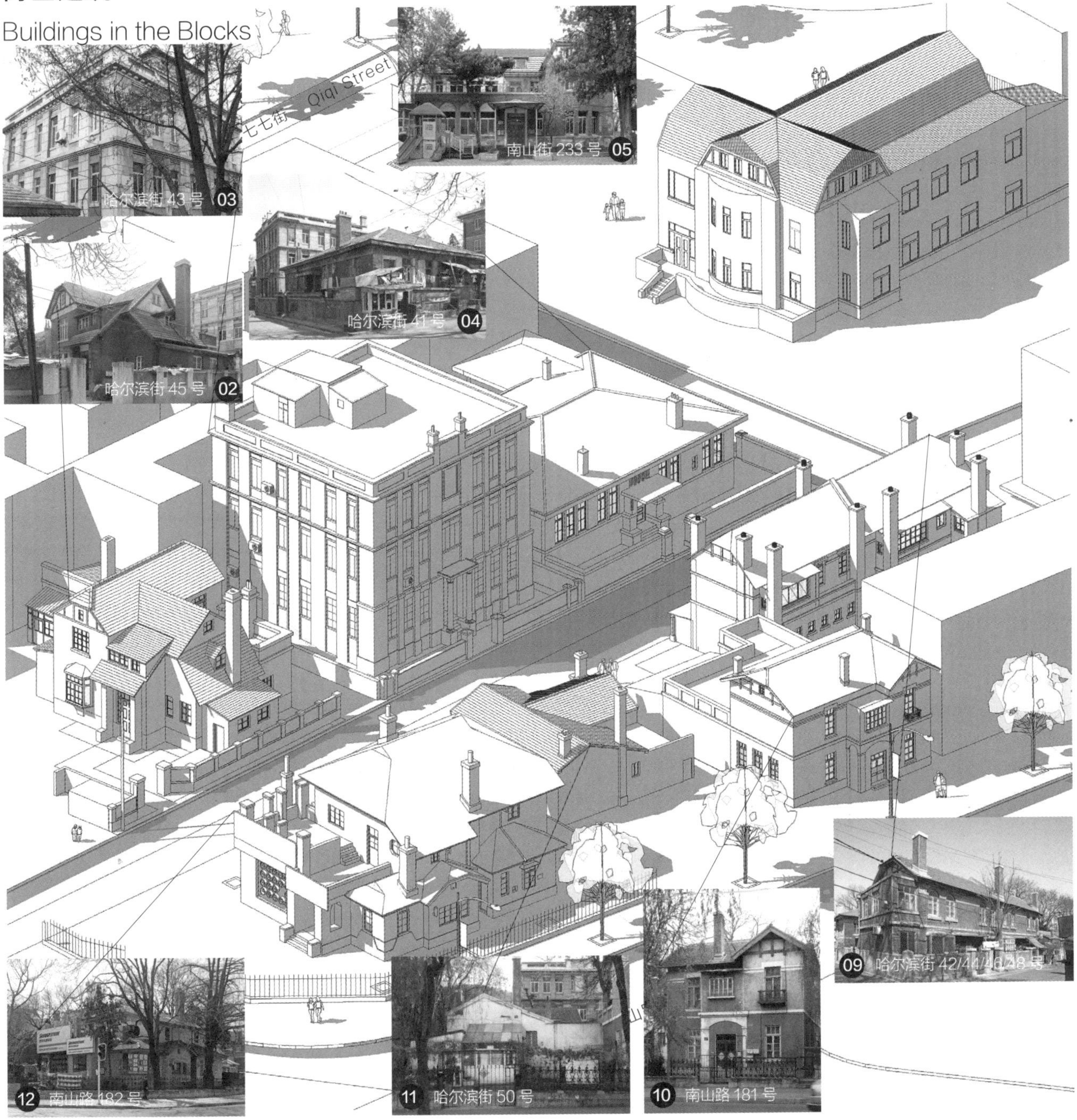

七七街 Qiqi Street
哈尔滨街 43 号 03
南山街 233 号 05
哈尔滨街 41 号 04
哈尔滨街 45 号 02
09 哈尔滨街 42/44/46/48 号
12 南山路 182 号
11 哈尔滨街 50 号
10 南山路 181 号

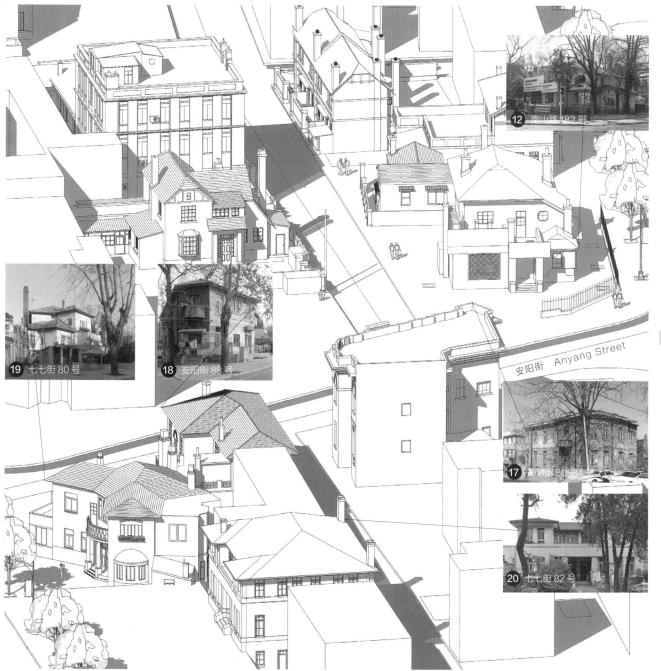

安阳街　Anyang Street

12　南山路182号

19　七七街80号

18　安阳街88号

17　安阳街80-84号

20　七七街82号

120

26 南山路215号

南山路 Nanshan Road

哈尔滨街76号 29

街102号 32

33 七七街104号

七七街108号 34

七七街110号 35

七七街 Qiqi Street

31 七七街100号

33 七七街104号

七七街

121

32 七七街102号

30 七七街98-1号

36 明泽街88号

37 枫林街36号

明泽街 Mingze Street

著者介绍 / Author

范悦

1988 年毕业于东南大学建筑系，1999 年获日本东京大学博士学位。现任深圳大学特聘教授、建筑与城市规划学院院长。曾任大连理工大学教授、建筑与艺术学院院长等。入选教育部新世纪优秀人才，获国家级教学果奖、2016 中国建筑设计奖·建筑教育奖等。

崔光勋

深圳大学建筑与城市规划学院助理教授、国家一级注册建筑师。

周博

大连理工大学建筑与艺术学院教授、博士生导师、人居环境研究所所长。

索健

大连理工大学建筑与艺术学院教授、硕士生导师。

王翔

大连理工大学 / 新加坡国立大学联合培养博士研究生。

本书的出版受国家自然科学基金重点项目"北方既有住区建筑品质提升与低碳改造的基础理论与优化方法"（51638003）的资助和支持。